Advances in Geoinformatics

Advances in Geoinformatics

Edited by Noel Lane

SYRAWOOD
PUBLISHING HOUSE

New York

Published by Syrawood Publishing House,
750 Third Avenue, 9th Floor,
New York, NY 10017, USA
www.syrawoodpublishinghouse.com

Advances in Geoinformatics
Edited by Noel Lane

International Standard Book Number: 978-1-68286-789-1 (Hardback)

Cataloging-in-Publication Data

Advances in geoinformatics / edited by Noel Lane.
    p. cm.
Includes bibliographical references and index.
ISBN 978-1-68286-789-1
1. Geoinformatics. 2. Geographic information systems.
3. Earth sciences--Data processing. I. Lane, Noel.
G70.212 .A38 2019
910.285--dc23

TABLE OF CONTENTS

Preface..**VII**

Chapter 1   **Full texts in the Czech geographical bibliography database**..**1**
           Eva Novotná

Chapter 2   **Accuracy evaluation of pendulum gravity measurements of Robert Daublebsky von Sterneck**...**7**
           Alena Pešková and Jan Holešovský

Chapter 3   **Geometric documentation of underwater archaeological sites**.....................................**15**
           Eleni Diamanti, Andreas Georgopoulos and Fotini Vlachaki

Chapter 4   **Monitoring of a concrete roof using terrestrial laser scanning**....................................**26**
           Ján Erdélyi, Alojz Kopáčik, Ľubica Ilkovičová, Imrich Lipták and Pavol Kajánek

Chapter 5   **Quality parameters of digital aerial survey and airborne laser scanning covering the entire area of the Czech Republic**.....................................**32**
           Jiří Šíma

Chapter 6   **Development and testing of INSPIRE themes Addresses (AD) and Administrative Units (AU) managed by COSMC**..**43**
           Michal Med and Petr Souček

Chapter 7   **Geodetic surveying as a tool for discovering the prehistoric settlement in Sudan (the 6th Nile cataract)**..**50**
           Jan Pacina

Chapter 8   **Database for tropospheric product evaluations - implementation aspects**...**67**
           Jan Douša and Gabriel Győri

Chapter 9   **Geostatistical Methods in R**...**81**
           Adéla Volfová and Martin Šmejkal

Chapter 10 **Results of GPS reprocessing campaign (1996-2011) provided by Geodetic observatory Pecný**...**106**
           Jan Douša and Pavel Václavovic

Chapter 11 **Panoramic UAV Views for Landscape Heritage Analysis Integrated with Historical Maps Atlases**..**119**
           Raffaella Brumana, Daniela Oreni, Mario Alba, Luigi Barazzetti, Branka Cuca and Marco Scaioni

Chapter 12   **Large geospatial images discovery: metadata model and technological framework**..130
Lukáš Brůha

Chapter 13   **Introducing the new GRASS module g.infer for data-driven rule-based applications**..146
Peter Löwe

Chapter 14   **Reference Data as a Basis for National Spatial Data Infrastructure**.........................157
Tomáš Mildorf and Václav Čada

Chapter 15   **PostGIS-Based Heterogeneous Sensor Database Framework for the Sensor Observation Service**...168
Ikechukwu Maduako

Chapter 16   **Quantum GIS plugin for Czech cadastral data**..185
Anna Kratochvílová and Václav Petráš

Chapter 17   **Minimal Detectable Displacement Achievable by GPS-RTK in CZEPOS Network**..193
Martin Raška and Jiří Pospíšil

Permissions

List of Contributors

Index

PREFACE

Geoinformatics is a field of science and technology, which employs and develops information science infrastructure to understand, analyze and solve cartographic, geoscientific and geographic problems. Research in this field has enabled its use in global and local environmental, security and energy programs. Some of the focus areas of geoinformatics include the development of geospatial database, analyzing geoinformation by using geographic information systems and geovisualization, geospatial analysis and modeling among many others. Various advances in this field have expanded its applications in diverse areas including virtual globes, urban planning, environmental modeling, transport network planning and management, public health, oceanography, land-use management, aviation, telecommunications, archeological reconstruction, biodiversity conservation, etc. This book unravels the recent studies in the field of geoinformatics. It strives to provide a fair idea about this discipline and to help develop a better understanding of the latest advances within this field. Students, researchers, experts and all associated with this subject will benefit alike from this book.

The researches compiled throughout the book are authentic and of high quality, combining several disciplines and from very diverse regions from around the world. Drawing on the contributions of many researchers from diverse countries, the book's objective is to provide the readers with the latest achievements in the area of research. This book will surely be a source of knowledge to all interested and researching the field.

In the end, I would like to express my deep sense of gratitude to all the authors for meeting the set deadlines in completing and submitting their research chapters. I would also like to thank the publisher for the support offered to us throughout the course of the book. Finally, I extend my sincere thanks to my family for being a constant source of inspiration and encouragement.

Editor

Full texts in the Czech geographical bibliography database

Eva Novotná

Director of Map Collection Head of Geographical Library

Charles University in Prague Faculty of Science Albertov 6, 128 43 Praha 2

Abstract

Open access to the documents is one of the basic requirements of databases users. Czech Geographical Bibliography On-line provides access to 185,000 bibliographical records of Bohemical geographic and cartographic documents and to more than 30,000 full texts and objects. The access is provided through a connection from the permanent storage, the Digital University Repository or a URL address of the bibliographical record. The works in public domain can directly become accessible or it is necessary to conclude licence agreement with authors, their heirs or with the editors of periodicals. Full texts of 14 titles of professional periodicals, university thesis, employees´ monographs or anthologies and on-line publications are available. Digitised maps have been connected to the database since 2012. 5,500 of them are accessible from the database since the beginning of 2014. The database is an important source both for professionals and general public interested in geography and cartography.

Keywords: open access, GEOBIBLINE, geographical databases, cartographic documents

1. Introduction

Providing access to full texts and objects is nowadays one of the regular requirements of the information service users. The requirement of open access to information, so called open access (hereafter OA) was internationally declared several times in Budapest (2002), in Bethesda (2003) and finally in Berlin (2003) (BARTOŠEK, M., 2009). Charles University signed on to the Berlin declaration in 2013 (UK, 2013). Bibliographic database Czech Geographical Bibliography On-line[1] (hereafter GEOBIBLINE) that has been produced since 2008 primarily by the Geographical library of the Faculty of Science of Charles University in Prague (NOVOTNÁ, 2011) in cooperation with other libraries headed by the National Library of the Czech Republic allows to generate metadata and subsequently to make accessible full texts of geographic and cartographic expert articles, monographs, chapters of books and anthologies, but also maps and graphics (NOVOTNÁ, 2013).

The accessibility depends on many factors: copyright laws, financing the production of bibliography of articles, creating metadata, preparation and editing of full texts and pictures,

---

[1] http://www.geobibline.cz

sufficient space for data storage in the repository, provision of lifespan and migration of data, but also on technologies of access.

The first concern is the copyright law (NOVOTNÁ, VONDRÁKOVÁ, 2012). On the one hand the documents of which the propriety copyright laws are already in public domain can be accessible; on the other hand it is possible to conclude exclusive or non-exclusive license contracts with authors, their heirs or with administrators of copyright. The Geographical library of the Faculty of Science of Charles University has been using both ways. The bibliography of articles was funded from the project of the Czech Ministry of Culture in 2008-2011. Original articles were created in the Geographical library. Thanks to the Czech Ministry of Culture TEMAP[2] project (technologies for making the map collections accessible) there are professionally catalogued and made accessible primarily digitised cartographic documents but also graphics and full texts of the articles. Metadata are usually generated from bibliographic records saved in the format MARC 21[3]. Full texts and pictures are saved in the Digital University Repository but also in the local storage at the Faculty of Science of Charles University. The access to data is carried out through a database where the full texts and pictures are attached as external links.

2. GEOBIBLINE database

The GEOBIBLINE database describes 185,000 Bohemical documents and provides access to a total of over 30,000 full texts and objects with metadata (to 1st April, 2014). It belongs among the world´s largest non-commercial databases in the field of geography. (NOVOTNÁ, 2012) It was primarily formed for the bibliography of a wide scale of documents from the field of geography and cartography of the 20th and 21st century. It was later extended chronologically to 1450. In content it was extended by full texts and objects. To ensure OA the database is using both green and gold way as well as other options. The green way means free access allowed by the author, the gold way is opened by the publisher and usually it is paid by the author. After the five-year experience from the Czech environment it can be observed that making something accessible in OA is not always so easily definable. There is a whole range of differences both in the editorial policy and in the approach of the authors and their heirs. In principle, a very individual approach is always required. The connection is carried out through the permanent storage of the Faculty of Science of Charles University, if the license for the work was obtained or if it is a work in public domain. The second, but unreliable option is to link URL addresses into the bibliographical record. Both types of connected full texts will display to the user as a result of the search in the left part of the screen as external links. After clicking the link it is necessary to agree the information on copyright law, where the researcher agrees to use the work only for his own need. After that the full text or the object opens. It can again have several forms depending on the way of connecting to the system.

2.1. Full texts in the permanent storage

The database contains full texts of articles, monographs, chapters of books and anthologies, as well as statistics. License contracts are gradually concluded with the editors of magazines,

---

[2] http://www.temap.cz
[3] http://www.loc.gov/marc/bibliographic/ecbdhome.html

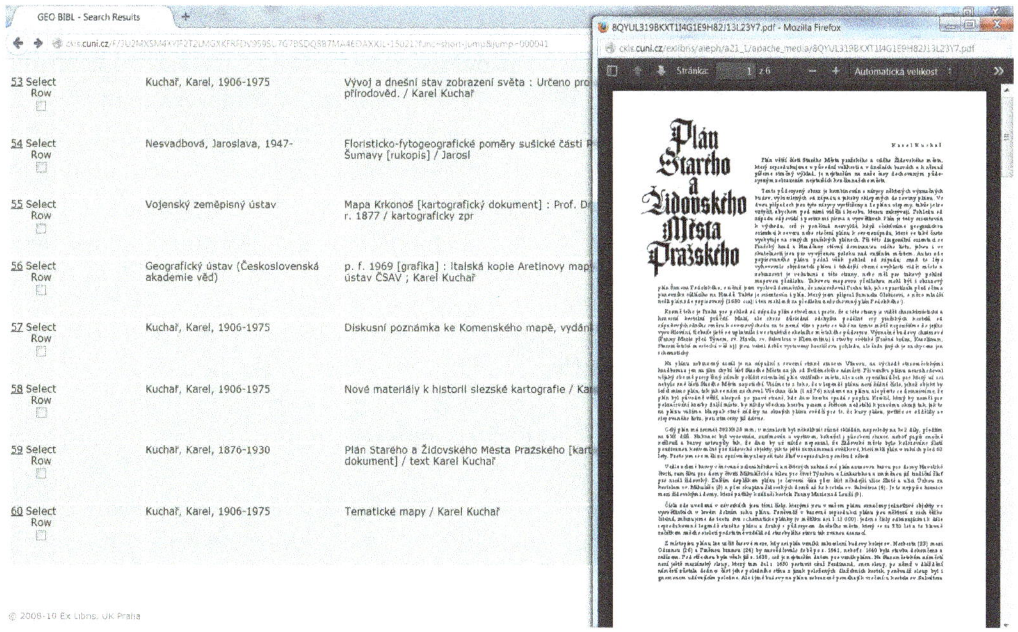

Figure 1: Example of the search interface where the external links to the full texts will display

each of which has its own publication policy. Sometimes the articles are published electronically before the printed release. It is led by an effort to speed up the process of transferring data to a potential interested person who may use and quote the article in the scientific research. Full texts of following titles are accessible: AUC Geographica, Acta Onomastica, Demografie, *Folia Facultatis scientiarum naturalium Universitatis Purkynianae Brunensis. Geographia, Geodetický a kartografický obzor,* Geografické rozhledy, Geografie, Kartografický přehled, Moravian geographical reports, Opera Corcontica, *Scripta Facultatis scientiarum naturalium Universitatis Purkynianae Brunensis. Geographia,* Sociologický časopis, Urbanismus a územní rozvoj, Vodní hospodářství, Vodohospodářské technicko-ekonomické informace and Vojenský geografický obzor. Detailed descriptions of the titles that are excerpted and supplemented by full texts can be found on the database website in the section Excerpted periodicals[4]. The amount of accessible volumes depends on the possibilities of the editors and of the Geographical library. In the last ten years the magazines have been editorially adapted from electronic data and the editors have usually archives available. The older articles are necessary to be scanned and read up by OCR software. The database excerpts 45 profile periodicals from the first volume. It then provides an access to 14 titles with full texts.

The accesses to monographs, chapters, anthologies are provided either directly by regular authors, i.e. employees of the Charles University or they are documents published on the Internet. The Czech Statistical Office data sets, that are commonly available electronically, are connected as well.

University theses are also a part of the database. Full texts have been accessible since 2010. License contracts on school works are concluded with students. GEOBIBLINE provides access to works from Masaryk University and Charles University.

---

[4] http://www.geobibline.cz/cs/node/29

License contracts are concluded with important authors or their heirs. The first contract was concluded with Prof. K. Kuchař, who provided license rights to the work of his father K. Kuchař. Professor Karel Kuchař (1906-1975) was an important representative of geography and cartography. Articles, monographs, reviews, New Year cards and reports have been digitized, described and added to the database. Some of the digitized documents were obtained through the contract from the Moravian library[5] project National Digital Library. A web portal devoted to Prof. Kuchař´s[6] work was created. 276 documents have been made accessible in the pdf format (CHRÁST, 2014).

A license contract with Eng. Dvořáček, the heir of Anna Dvořáčková who is a co-author of Kuchař´s book on the Moll map collection[7] in Brno was concluded in 2013.

2.2. Full texts from the linked URL addresses

Linking full texts published on the Internet represents the second way of access. This method is not ideal, particularly because of the lack of stability of addresses and different forms of publication that are problematic. The point is that the editors often publish whole volumes or issues, which slows down the search and it is not possible to connect them directly to a specific bibliographic record of the article. In spite of a regular control of internet addresses functionality the method is not reliable. However, there is no other option until the license contract is concluded.

2.3. Cartographic documents in the database

The database Czech Geographical Bibliography On-line makes also accessible cartographic documents, old maps, atlases and globes (until 1850). Altogether it contains 31,000 bibliographic records of such special documents. 23 of them originated in the 16th century, 123 in the 17th century, 776 in the 18th century, from the 19th century then entire 4546 records, from the 20th century 18432 and finally from the 21st century 7874 records.

The Charles University Computer Centre has prepared a script that lists newly imported objects to a file. The script lists PID, scans the barcode field identifier in the technical metadata. Afterwards it sends a query to Aleph and adds to the PID record system a number and a title of the map sheet (according to MARC 21 field) for better traceability. Thus a generated list is sent to the Map Collection of the Faculty of Science. Bibliographic records are then connected to the objects of maps and atlases through Z39.50 protocol based on the title from the generated list. (NOVOTNÁ, 2013). Searches for Bohemical documents are made for the GEOBIBLINE database and then they are connected to the database. It is possible to view maps, zoom in and out by clicking on an external link from the database catalogue. The maps are available in the University Repository[8] or directly from the GEOBIBLINE database in the jpeg2000 format at a resolution of 300 DPI under the icon FULLTEXT. The descriptive metadata can be searched in two interfaces (simple and advanced), according to fields: title, author (inverted), secondary authors, subject heading, publisher, year of publication, genre/form and language. The records may be also only browsed through and

---

[5] https://www.mzk.cz/o-knihovne/odborne-cinnosti/ndk
[6] http://web.natur.cuni.cz/gis/kuchar/
[7] http://mapy.mzk.cz/en/
[8] http://repositar.cuni.cz

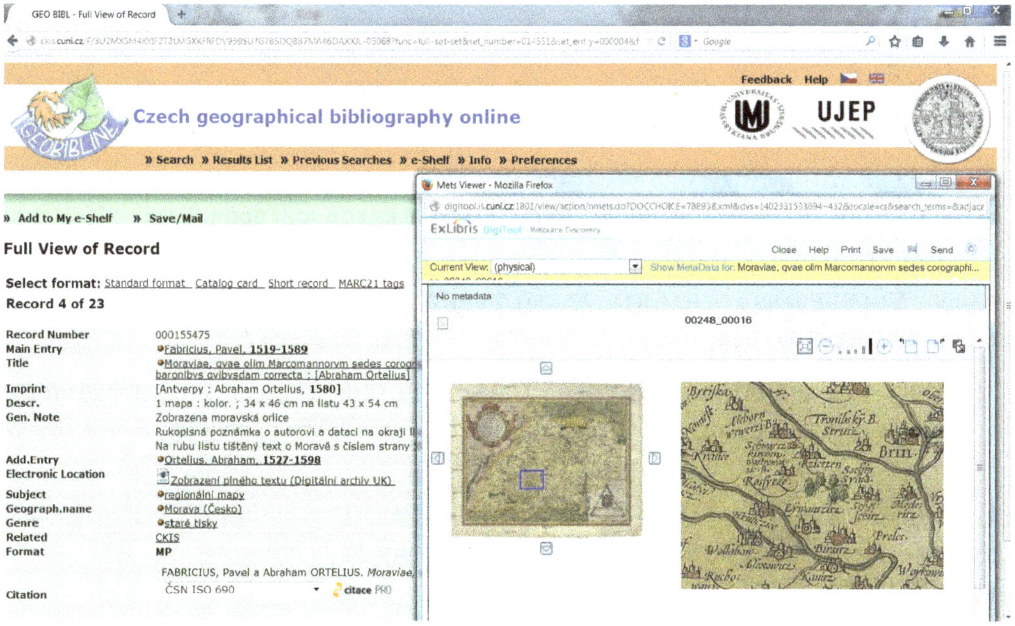

Figure 2: Sample of a map display from the GEOBIBLINE database

metadata can be viewed in three formats. The Help icon (a question mark) is up on the right. The access is possible both in Czech and English (globe icon). Metadata are in MIX format.

The GEOBIBLINE database provides an on-line access to 20 maps from the 16th century, 96 maps from the 17th century, 689 maps from the 18th century and 4695 from the 19th and 20th centuries.

3. Conclusion

The database development continues thanks to the TEMAP project. A shared collection of bibliographic records from participating libraries and original cataloguing of the works in the Geographical library and in the Map Collection of the Faculty of Science are going on. Digitized maps are also gradually being uploaded to the database and connected to the records. The digital repository contains 32,000 digitized cartographic documents. By the end of the project in 2015 their number should increase of 20,000. Not all of them are, however, cartographic bohemics. Even so, their share will be fairly high. Photographs of globes and telluriums are beginning to be connected as well. The database websites collect also statistical information on accesses from abroad. Apart from the Czech users the websites are frequently visited from the United States of America, Germany, the Netherlands, Russia, Ukraine and Spain. On the contrary, the number of visitors from Slovakia is surprisingly low.

Acknowledgement

This article was supported through the project of the Czech Ministry of Culture, DF11P01O-VV003, TEMAP – Technology for discovering of map collections of the Czech Republic: methodology and software for protection and use of cartographic works of the national cartographic heritage.

References

[1] BARTOŠEK, M. 2009. *Open access - otevřený přístup k vědeckým informacím. Úvod do problematiky. Zpravodaj ÚVT MU. ISSN 1212-0901, 2009, roč. XX, č. 2, s. 1-7.*

[2] CHRÁST, J. 2014. Nový web věnovaný prof. Karlu Kuchařovi. Ikaros [online]. 2014, roč. 18, č. 2 [cit. 04.04.2014]. Available: `http://www.ikaros.cz/node/8155`

[3] NOVOTNÁ, Eva a kolektiv. 2011. *Geografická bibliografie ČR online: GEOBIBLINE.* Praha : VŠCHT. 152 s. : il. ISBN 978-80-7080-773-6. Available: `http://vydavatelstvi.vscht.cz/katalog/uid_isbn-978-80-7080-773-6/anotace/`.

[4] NOVOTNÁ, Eva a Alena VONDRÁKOVÁ. 2012. Zpřístupnění a užití digitalizovaných kartografických děl. Geografické rozhledy. 2012, roč. 22, č. 3, příloha, s. 1-4. ISSN 1210-3004.

[5] NOVOTNÁ, Eva. 2012. Base de datos GEOBIBLINE y su comparación con las bases geográficas existentes en el mundo. El profesional de la información. Roč. 21, č. 3, s. 304-311. ISSN 1386-6710.

[6] NOVOTNÁ, Eva. 2013. Staré mapy a grafiky v Geografické bibliografii ČR on-line. Knihovna - knihovnická revue [online]. Roč. 24, č. 1, s. 5-27. ISSN 1801-3252. Available: `http://knihovna.nkp.cz/knihovna131/13105.htm`

[7] *Univerzita Karlova. 2013.* UK se přihlásila k Berlínské deklaraci Open Access. Iforum [online]. 2013, [cit. 04.04.2014]. Available: `http://iforum.cuni.cz/IFORUM-14799.html`

Accuracy evaluation of pendulum gravity measurements of Robert Daublebsky von Sterneck

Alena Pešková, Jan Holešovský

Department of Geomatics, Faculty of Civil Engineering
Czech Technical University in Prague
Thákurova 7, 166 29 Prague 6, Czech Republic
alena.peskova@fsv.cvut.cz, jan.holesovsky@fsv.cvut.cz

Abstract

The accuracy of first pendulum gravity measurements in the Czech territory was determined using both original surveying notebooks of Robert Daublebsky von Sterneck and modern technologies. Since more accurate methods are used for gravity measurements nowadays, the work [3] is mostly important from the historical point of view. In previous works [5], the accuracy of Sterneck's gravity measurements was determined using only a small dataset. Here we process all Sterneck's measurements from the Czech territory (a dataset ten times larger than in the previous works [5]), and we complexly assess the accuracy of these measurements. Locations of the measurements were found with the help of original notebooks. Gravity in the site was interpolated using gravity model EGM08, resultant gravity is in actual system S–Gr10. Finally, the accuracy of Sterneck's measurements was evaluated on the base of the differences between the measured and interpolated gravity.

Keywords: Robert Daublebsky von Sterneck, relative pendulum measurements, gravity.

1. Introduction

Robert Daublebsky von Sterneck (* 7.2.1839, † 2.11.1910) was born in Prague, he acted as geodesist, astronomer and geophysicist. He was Head of Astronomical Observatory Institute in Vienna in 1880-1884 and also he was the first to make gravimetrical measurements in the Austria-Hungary. Although he worked in army all his life, he also did various surveying and astronomical measurements. His work was recognized and his name is famous also nowadays. The pendulum instrument built by Sterneck himself was used for gravity measurements, and its improved version was also used in other countries in Europe. Daublebsky used a relative method to measure gravity. Only the time of swing of the pendulum was measured with four implemented corrections. The initial gravity point was located in the cellar of Military Geographical Institute in Vienna, with value g = 980 876 mGal [2]. We divided Sterneck's measurements to two datasets. The first rule for division was different localities of the measurements (measurements on hilltops near trigonometrical points, and measurements in buildings in towns). The second rule was the time of measurements (there is a 3 year gap between the two datasets).

2. Localization of Sterneck's gravity measurements

The original Daublebsky's surveying notebooks [6] and a summary of results in technical report [4] were used for gravity measurement localization. The technical report contains approximate astronomical coordinates of the measurements, whereas detailed information about the measurement process and locations is given in the notebooks. From the technical report, were used these informations: year of measurement, number and title of point (Czech and Germany), latitude and longitude, elevation and the measured value of gravity. Only the details about the locations were used from the notebooks. These details were not registered for all measured points, - 15 points measured in towns haven't had any information about their location (these points were locallized only by approximate coordinates and heights). The measurements were divided into two groups: both by measurement location and by the time measurement. In 1889 – 1895, - 106 points were determined in the Czech territory, as is shown in Figure 1. The first group of points is located on hilltops close to know trigonometric points – hilltop dataset (blue circles in Figure 1). In 1889 – 1891 were determined 35 points in close trigonometric points and 6 points with differently locations in the Czech territory. In 1894 – 1895 (after a 3 year gap), the second group of 65 points was measured in buildings inside towns in the Moravian territory – building dataset (green squares in Figure 1).

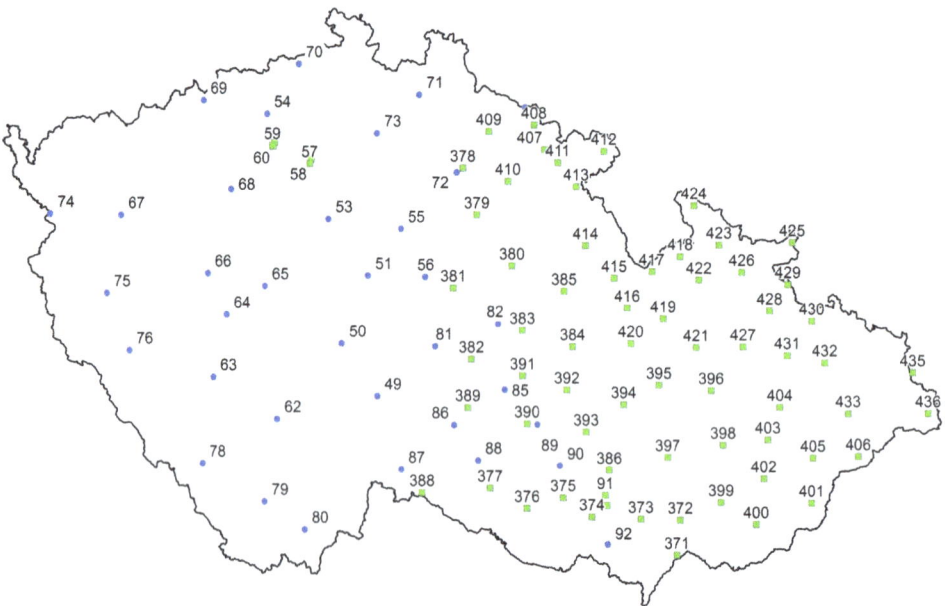

Figure 1: Locations of Sterneck's gravity measurements.

3. Determination of the gravity differences

We used the ArcMap program to determine the coordinates of the measurements with joint WMS provided by The Czech Office for Surveying, Mapping and Cadastre (ČÚZK). Coordinates of the locations with error estimates and corrections for heights (e.g. measurement in a building or on top of a lookout tower) was provided by The Department of Gravimetry, Land Survey Office (ZÚ). They intepolated the complete Bouguer anomaly using the methods of ordinary kriging. The results of interpolation are the most probable values of gravity for the referenced locations, given with their upper and lower estimate limits. The gravity value is found in this interval with 95% probability. The limits are affected by the uncertainty in elevation and position. The estimated interval isn't symmetrical and it is different for each of the measured points. Throughout this work, only the most probable gravity values were used. The gravity differences are calculated as the difference between Sterneck's measured gravity and the interpolated gravity. These differences were used to evaluate the accuracy of Daublebsky's pendulum gravity measurements.

4. Data analysis

The differences between the measured and interpolated gravity values are distinctly different for the hilltop and the building dataset. The differences gravity in the building dataset show a systematical offset +21.7 mGal, shown in Figure 2. This displacement represents a 72 meters error in elevation. The cause of this displacement isn't known, therefore both datasets were processed separately. A surprising fact about building dataset is that the gravity differences for points without precise location information (only approximate coordinates and heights) and points with these information weren't significantly different. This is illustrated in Figure 3.

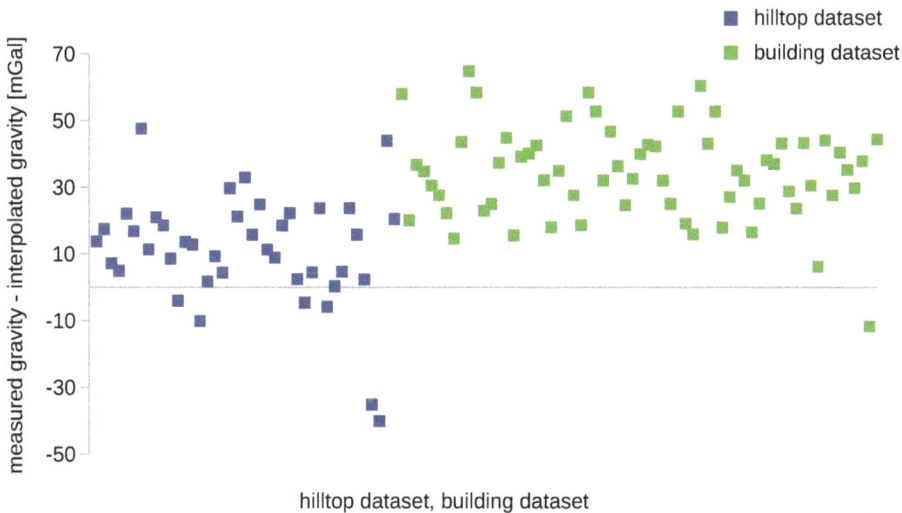

Figure 2: Differences between measured and interpolated gravity for both datasets.

The datasets were tested for data quality. Dependencies between various quantities were tested for this purpose using hypothesis verification. The computed correlation coefficient was compared with its critical value. The tested hypotheses are: gravity falls with growing elevation – (H1), gravity grows with growing latitude – (H2), and gravity and longitude are independent – (H3). All three hypotheses were verified for the hilltop dataset. In the building dataset, H2 and H3 were also verified, but H1 not. Because all of the tested quantities in the building dataset are all right, we think that the elevation values are also affected by an error different from Gaussian noise. Still, the building dataset was used in other processing.

The accuracy of Sterneck's measurements was evaluated by several methods. First, we deter-

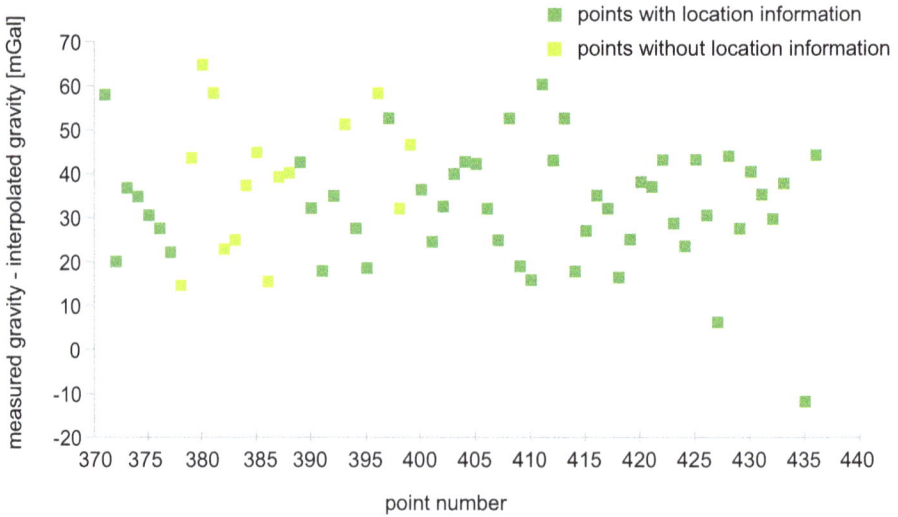

Figure 3: Differences between measured and interpolated gravity for the building dataset.

mined the value mean gravity difference. This value shows the magnitude of the difference
between Sterneck's measured gravity and the interpolated gravity. The hilltop dataset has
mean gravity difference +11.6 mGal, and +33.3 mGal is for the building dataset. These means
are apparently affected by an unknown displacement of the used gravity systems. However,
the computed differences are only valid with the assumption of null displacement between the
gravity systems. If we want to compare the accuracy of the past and recent measurements,
we must calculate the mean difference from absolute value of gravity difference. The datasets
are characterized by the mean absolute value of gravity difference of 12.9 mGal for the hilltop dataset and 33.3 mGal for the building dataset. The second method is to evaluate the
precision of the measurements using standard deviation of the mean gravity difference. This
value shows precision of the measurement method and removes the systematic displacement
between the two datasets. Both datasets have identical value of standard deviation equal
to 10.3 mGal. The conclusion is that both datasets have identical measurement accuracy,
although they were determined with different conditions and in a different environment.

5. Discussion and conclusion

Sterneck's measurements were divided into two dataset differing by both the type of the measurement locations and the time of their acquisition. The statistical processing and evaluation
was done separately because of these differences. The building dataset is displaced system-
atically by about +21.7 mGal from the hilltop dataset (mean gravity difference 11.6 mGal
for the hilltop dataset and 33.3 mGal for the building dataset). The cause of this systematic
displacement is unknown. The building dataset was determined after 3 year gap. During
this time some parameters of the pendulum instrument or some changes in way of calculating
corrections could be changed. These changes probably can cause the systematic displacement between both of datasets. Therefore the accuracy of Sterneck's measurements is better
assessed by standard deviation of the mean difference. That is 10.3 mGal and is identical for both datasets. This value can be compared with Sterneck's precision estimate of

10 mGal [4]. The mean of gravity difference 11.6 mGal for the hilltop dataset and 33.3 mGal for the building dataset can be compared to measurements in Hungary where the errors of Sterneck's measurements are up to ± 20 mGal [5], (but the difference for some points is up to 25 mGal [1]).

Acknowledgement

The authors thank the employees of The Department of Gravimetry, Land Survey Office (ZÚ) in Prague; Martin Lederer, who borrowed the original surveying notebooks of Robert Daublebsky von Sterneck and Otakar Nesvadba, who interpolated the gravity values.

References

[1] Alexandr Drbal and Milan Kocáb. "Významný rakouský generálmajor Dr.h.c. Robert Daublebsky von Sterneck". In: *Geodetický a kartografický obzor* 56(98).2 (2010), pp. 40–46. URL: http://archivnimapy.cuzk.cz/zemvest/cisla/Rok201002.pdf.

[2] Martin Lederer. "Historie kyvadlových měření na území České republiky". In: *Geodetický a kartografický obzor* 58(100).6 (2012), pp. 129–133. URL: http://archivnimapy.cuzk.cz/zemvest/cisla/Rok201206.pdf.

[3] Alena Pešková. "Hodnocení přesnosti kyvadlových tíhových měření R. Sternecka". Master thesis. Czech Technival University in Prague, 2015. URL: http://geo.fsv.cvut.cz/proj/dp/2015/alena-peskova-dp-2015.pdf.

[4] Zdeněk Šimon. *Kyvadlová měření v letech 1956 - 1962*. Tech. rep. Geodetický a topografický ústav v Praze, 1962.

[5] V. B. Staněk and J. Potoček. "Vývoj a způsob měření intensity tíže v Čechách a na Moravě". In: *Zeměměřičský obzor* 1(28).6 (1940), pp. 81–87.

[6] Robert Sterneck. "Měřické sešity 1889 - 1895". Vojenský zeměpisný ústav ve Vídni. Unpublished.

Table 1: Input – Part 1

Year of measurement	Number of point	Latitude [° ']	Longitude from Ferro [° ']	Altitude [m]	Measured gravity [mGal]
1889	49	49 24	32 38	738	980 856
	50	49 36	32 20	712	980 887
	51	49 55	32 27	545	980 938
	52	50 44	33 24	1602	980 762
	53	50 08	32 08	356	981 016
	54	50 33	31 36	835	980 924
	55	50 08	32 39	213	981 070
	56	49 57	32 51	470	980 952
	57	50 22	31 57	205	981 076
	58	50 23	31 57	459	981 019
	59	50 25	31 40	202	981 060
	60	50 26	31 41	417	980 998
	61	50 25	31 41	250	981 055
1890	62	49 14	31 58	624	980 846
	63	49 22	31 29	585	980 851
	64	49 39	31 31	842	980 855
	65	49 48	31 45	659	980 911
	66	49 49	31 20	716	980 893
	67	50 01	30 40	822	980 922
	68	50 12	31 25	534	980 983
	69	50 34	31 08	921	980 920
	70	50 48	31 47	748	980 963
	71	50 44	32 39	1010	980 915
	72	50 25	32 59	430	981 016
	73	50 32	32 23	565	980 989
	74	49 58	30 10	939	980 862
	75	49 40	30 39	537	980 937
	76	49 26	30 52	724	980 877
	78	49 00	31 29	1362	980 663
	79	48 52	31 57	1084	980 716
	80	48 46	32 15	869	980 760
1890	81	49 39	32 59	709	980 849
	82	49 47	33 24	662	980 895
1891	85	49 30	33 30	693	980 881
	86	49 19	33 11	732	980 873
	87	49 05	32 51	731	980 819
	88	49 10	33 22	710	980 861
	89	49 22	33 45	639	980 841
	90	49 11	33 56	513	980 846
	91	49 05	34 16	201	981 004
	92	48 52.0	34 19.0	550	980 853

Table 2: Input – Part 2

Year of measurement	Number of point	Latitude [° ']	Longitude from Ferro [° ']	Altitude [m]	Measured gravity [mGal]
1894	371	48 51.3	34 47.7	160	980 943
	372	49 00.6	34 47.8	193	980 917
	373	48 59.7	34 31.5	226	980 943
	374	48 58.9	34 11.3	181	980 957
	375	49 03.0	33 58.8	246	980 961
	376	48 59.1	33 44.5	355	980 937
	377	49 03.3	33 28.5	465	980 925
1895	378	50 26.3	33 01.3	273	981 057
	379	50 14.5	33 09.5	228	981 068
	380	50 02.3	33 26.8	214	981 076
	381	49 54.6	33 03.5	263	981 054
	382	49 36.5	33 14.7	428	980 946
	383	49 45.7	33 34.3	569	980 935
	378	50 26.3	33 01.3	273	981 057
	379	50 14.5	33 09.5	228	981 068
	380	50 02.3	33 26.8	214	981 076
	381	49 54.6	33 03.5	263	981 054
	382	49 36.5	33 14.7	428	980 946
	383	49 45.7	33 34.3	569	980 935
	384	49 42.9	33 55.9	555	980 955
	385	49 57.3	33 49.7	287	981 030
	386	49 11.7	34 16.5	235	980 962
	387	49 02.3	34 17.1	191	980 979
	388	48 59.9	33 01.0	506	980 911
1895	389	49 23.7	33 15.5	514	980 940
	390	49 21.3	33 40.7	425	980 955
	391	49 33.7	33 36.6	574	980 922
	392	49 31.4	33 55.5	554	980 942
	393	49 21.0	34 05.3	270	980 999
	394	49 29.3	34 19.7	396	980 969
	395	49 35.4	34 33.3	410	980 953
	396	49 35.4	34 55.3	225	981 026
	397	49 16.7	34 40.0	254	981 001
	398	49 21.5	35 02.3	200	980 983
	399	49 06.3	35 03.7	209	980 958
	400	49 01.4	35 18.8	248	980 932
	401	49 08.4	35 40.7	390	980 892
	402	49 13.7	35 20.2	231	980 959
	403	49 24.0	35 20.5	316	980 972
	404	49 32.9	35 24.2	256	981 010
	405	49 20.4	35 39.7	340	980 954

Table 3: Input – Part 3

Year of measurement	Number of point	Latitude [° ']	Longitude from Ferro [° ']	Altitude [m]	Measured gravity [mGal]
1895	406	49 21.9	35 58.5	510	980 906
	407	50 33.8	33 34.9	415	981 052
	408	50 39.8	33 29.1	610	981 045
	409	50 36.7	33 10.4	462	981 052
	410	50 24.3	33 21.0	335	981 039
	411	50 30.8	33 41.0	359	981 097
	412	50 35.2	33 59.9	405	981 085
	413	50 25.1	33 49.8	337	981 069
	414	50 09.9	33 56.6	321	981 014
	415	50 02.2	34 10.0	368	981 007
	416	49 54.8	34 16.8	387	981 002
	417	50 05.1	34 25.6	567	980 972
	418	50 09.8	34 36.8	536	980 969
	419	49 53.0	34 32.3	301	981 000
	420	49 45.5	34 19.9	350	981 002
	421	49 46.3	34 47.3	235	981 025
	422	50 04.2	34 45.6	489	981 005
	423	50 13.9	34 52.5	441	981 023
	424	50 23.5	34 40.4	339	981 043
	425	50 16.5	35 22.9	238	981 081
	426	50 07.4	35 03.1	519	981 003
	427	49 47.7	35 06.6	550	980 944
	428	49 58.0	35 16.2	550	980 999
	429	50 05.4	35 22.7	313	981 041
	433	49 32.9	35 52.9	406	980 973
	435	49 45.1	36 18.3	308	980 972
	436	49 34.7	36 26.0	386	980 973

Geometric documentation of underwater archaeological sites

Eleni Diamanti[1], Andreas Georgopoulos[1], Fotini Vlachaki[2]

[1]Laboratory of Photogrammetry, National Technical University of Athens
drag@central.ntua.gr

[2]Hellenic Institute of Marine Archaeology
ienae@otenet.gr

Abstract

Photogrammetry has often been the most preferable method for the geometric documentation of monuments, especially in cases of highly complex objects, of high accuracy and quality requirements and, of course, budget, time or accessibility limitations. Such limitations, requirements and complexities are undoubtedly features of the highly challenging task of surveying an underwater archaeological site. This paper is focused on the case of a Hellenistic shipwreck found in Greece at the Southern Euboean gulf, 40-47 meters below the sea surface. Underwater photogrammetry was chosen as the ideal solution for the detailed and accurate mapping of a shipwreck located in an environment with limited accessibility. There are time limitations when diving at these depths so it is essential that the data collection time is kept as short as possible. This makes custom surveying techniques rather impossible to apply. However, with the growing use of consumer cameras and photogrammetric software, this application is becoming easier, thus benefiting a wide variety of underwater sites. Utilizing cameras for underwater photogrammetry though, poses some crucial modeling problems, due to the refraction effect and further additional parameters which have to be co-estimated [1]. The applied method involved an underwater calibration of the camera as well as conventional field survey measurements in order to establish a reference frame. The application of a three-dimensional trilateration using common tape measures was chosen for this reason. Among the software that was used for surveying and photogrammetry processing, were Site Recorder SE, Eos Systems Photomodeler, ZI's SSK and Rhinoceros. The underwater archaeological research at the Southern Euboean gulf is a continuing project carried out by the Hellenic Institute for Marine Archaeology (H.I.M.A.) in collaboration with the Greek Ephorate of Underwater Antiquities, under the direction of the archaeologist G.Koutsouflakis. The geometric documentation of the shipwreck was the result of the collaboration between H.I.M.A. and the National Technical University of Athens.

Keywords: underwater photogrammetry, trilateration, underwater camera calibration, visualization, orthophotomosaics, 3D reconstruction, ancient shipwreck

1. Introduction

With 17,000 kilometers of coastline, equivalent to 25% of the total Mediterranean coast, with almost 3,500 islands and at least 1,000 shipwrecks detected in the Greek seas, Greece is a country with one of the largest and perhaps the most important underwater archaeological

heritage. Ancient shipwrecks, submerged settlements or ancient harbors are housed for centuries in the Greek seas. Nevertheless, practical implication of theoretical and technological
developments in the fields of surveying and photogrammetry for the geometric documentation of this underwater heritage is far behind the rapid developments and innovations that
are applied when it comes to surveying 'terrestrial' monuments. This paper presents an effort to improve and experiment on the synergy of conventional surveying techniques, such
as simple tape measurements and trilateration adjustments, with modern software and digital technology, in order to produce 3 dimensional reconstructions that can assist underwater
archaeologists in reaching their scientific conclusions, through a geometrically accurate, documentation of the site and the excavation process, as well as to bring those who do not have the
opportunity to access this submerged monument, on a 'digital trip' to an ancient shipwreck
in deep waters.

1.1. Description of the object

The Hellenistic shipwreck, whose case is examined in this paper, was found in 2006 in the
northwest side of the island Styronisi at Southern Euboean gulf, at a depth range of depth
between 39 and 47 meters below the sea surface. The shipwreck dates back to the Late
Hellenistic period (late 2nd to early half of the 1st century B.C.) and it is the only ancient
shipwreck that was detected in Southern Euboean gulf, in such a relatively good condition.
The dimensions of the exposed shipwreck are approximately 18 meters long and 7 meters
wide. The cargo of the ship consists, mainly, of intact and broken amphorae, 90% of which
are considered as Brindisi type of amphorae. Additionally, among the ship's cargo interesting
objects were found, such as parts of luxurious bronze furniture, bronze and steel spikes, a stone
wash basin, parts of the harness of the ship and broken tiles. Among the most important finds
of the whole archaeological survey, is a small part of the dress of a natural size bronze statue
and beneath the surface layer of sand, two parts of the wooden hull of the vessel, something
very rare, since wood cannot be preserved for such a long time in underwater archaeological
sites.

Since the discovery of the specific wreck is considered of great importance, H.I.M.A. and
the supervising archaeologist intended to launch a systematic investigation and excavation of
this monument. Therefore, the detailed and accurate documentation of the site became an
immediate priority. Orthophotomosaics and 3D rendered models of the wreck were considered
as the ideal products, in order to map the site in the condition that it was found and prior
to excavation [2]. Of course, when it comes to the excavation period, the requirements are
increasing; daily recording of the excavation trenches, mapping of the 3D locations of the
artifacts, 3D reconstruction of the shipwreck excavation, production of daily 2D plans or
3D measurement and modelling of finds are some of the needs that arise for the complete
documentation of such a monument.

1.2. Underwater surveying & underwater photogrammetry

It is well-known that conventional mapping is a process subject to human error in underwater archaeology [3, 4, 5], while photogrammetry has long been a viable technique in such
situations [6]. Thus the main objective is to create a 3-dimensional model of the site using photogrammetry, which could be dynamically updated in the future according to the
progress of the ongoing archaeological excavation. Underwater photogrammetry clearly offers

some advantages for the surveying of a submerged site thus overcoming difficulties such as limited on-site accessibility and non-destructive efficiency. On the other hand, some crucial and inevitable matters that have to be faced and co-estimated arise in such conditions: no operational control on data acquisition, low image quality caused by poor underwater lighting (e.g. variations of scattering or absorption of red wavelength especially in deep waters) even if artificial lighting is used, two-media (air-water) data collection, , significant diffusion that complicates the object recognition and the tie point measurements and, last but not least, control point establishment limitations as common tape measurements and 3D trilateration methods are perhaps the only plausible methods. Despite all the difficulties encountered, photogrammetric software can be increasingly extended from land-based applications to underwater applications.

2. The geometric documentation of the hellenistic shipwreck in south Euboea

2.1. Establishing an underwater control points network

One of the main tasks of the surveying procedure on site was to establish an underwater network of control points, which would be measured and calculated with tape trilateration adjustment. Once the theoretical control network had been designed in terms of adequate geometry, i.e. widely dispersed control points and efficient stability of the points, it had to be set up on the site. It is at this point, where the problems associated with surveying under water start to affect the quality of the survey. The tape survey is dependent on the divers' ability to install control points in geometrically correct positions of high rigidness, as well as to measure the distances between the points with sufficient accuracy. In this case, 20 control points were established. They were made of targets stuck on 10x10cm2 Plexiglas tiles; fourteen of them, fitted onto 0.5m long metal rods, were inserted in sand as deep as possible (Fig.1a) and the remaining six were fitted, with common tie-wraps, on the mouths of 6 intact amphorae (Fig.1b). The tiles bearing the target points were also labeled with numbers, so that they could be easily identified by the divers.

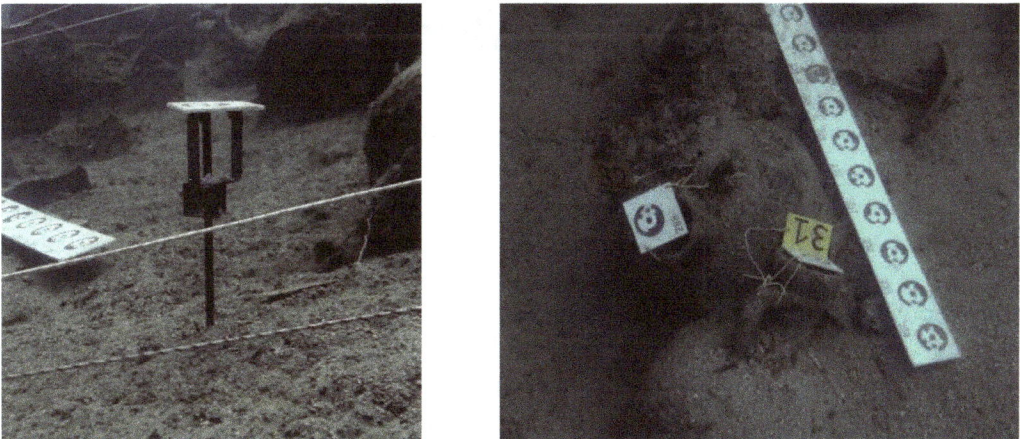

Figure 1: (a) Control points inserted in sand, (b) Control points fitted on the mouth of amphorae

The position of each control point was measured from at least 5 other control points of the network. From a total number of 119 measured distances, 82 were selected and adjusted using the Site Recorder SE software through a large least-squares network adjustment. In order to obtain three-dimensional coordinates for a point, the minimum number of measurements is of course three. With three measurements, error or reliability cannot be estimated. Therefore, at least 5 measurements from each point were taken, so that the accuracy of the coordinates of each control point could be estimated. The total RMS of the trilateration adjustment resulted to 0.027m.

2.2. Data acquisition

For the optimal organization of the photogrammetric restitution, a photomosaic of the shipwreck was created as an approximate complete mapping of the site. The Hugin open source software, a piece of software typically used for stitching and blending a series of photographs was employed. This software was developed by B. Hartzler, an archaeologist and member of H.I.M.A. [7]. This first photomosaic proved to be a very useful tool for the photogrammetric image data acquisition that followed.

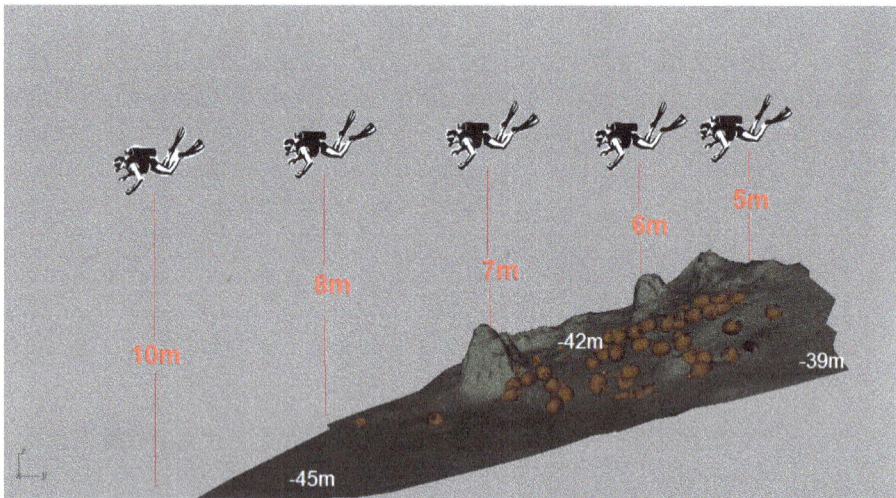

Figure 2: Photo data capture

For the image data acquisition, a SONY DSLR-A700 camera and an Ikelite® protecting housing were available. The camera has a 12-mm lens and all individual images are of a resolution of 4272x2848 pixels. The diver-photographer "flew" over the wreck (Figure 2), taking about 120 photos in 4 strips, with a 70-80% forward overlap and 50% side overlap. The physics of underwater light diffusion requires that images should be taken as close to the object as possible. Images taken from longer distances have, as a result, much lower quality. A standard "flying height", a strictly straight strip line and a satisfying overlap between images are definitely requirements of an optimal data acquisition for photogrammetric processing purposes. Nevertheless, it seems to be a really challenging task when photographing under such conditions. In this case, the limited available time of the archaeological research, the increased difficulty to approach the site and the strong underwater currents did not favor the image acquisition process with the aforementioned requirements.

3. Photogrammetric processing

3.1. Underwater camera calibration

The accurate 3D reconstruction, as well as the pose estimation of an object, from images, requires the thorough knowledge of the intrinsic camera parameters i.e. focal length, principal point's coordinates and lens distortion. The underwater camera calibration problem has been treated in several ways so far. A 'standard case' of multimedia photogrammetry [8]: three media; an object in water, a sensor located in air and a transparent plane of the camera housing separating the object from the sensor. As far as using images for underwater surveying is concerned, there are generally two categories for approaching camera calibration; the first is based on dry camera calibration methods, where the intrinsic parameters of a camera immersed in water, or any other fluid, can be calculated from an air calibration, as long as the optical surface between the two fluids presents some simple geometrical properties [9]. The second approach, is not based on modeling any parameters that have to do with different media through mathematical models, but treats the system camera-housing (or air-glass-water) as a unique system.

The camera calibration in this project's case is based on the second approach and was carried out using the Photomodeler® calibration module. A SONY DSLR A700 camera, in a water-proof Ikelite housing device, was immersed into water and a total number of 24 images of the board bearing the calibration pattern, i.e. a grid of specific dots (Figures 3a, b, c) were taken. Photomodeler, firstly, analyses each picture using a line interpolation algorithm to find and mark the dots and the 4 control points of the plane pattern [10]. Seventeen (17) images were processed into an image processing software in such a way that the algorithm could detect only the coded dots-targets (Figure 3b). The scale was constrained by the 4 known distances between the control points of the pattern.

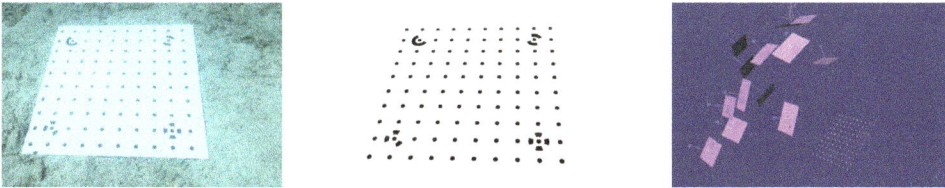

Figure 3: (a) Initial calibration image, (b) Corrected calibration image, (c) Camera positions.

On completion of the calibration, the intrinsic parameters of the camera were determined, including the principal point's coordinates, the radial distortion values and a focal length of 15.03mm. As far as the focal length is concerned, the ratio computed-to-nominal value (15,03mm/13mm) was found to be 1.16. When compared to the refractive index (1.34), it is obvious that underwater refraction is not the only parameter that has to be estimated, contrary to the dry camera calibration procedures. Depth, temperature and salinity can be considered as unstable parameters affecting a typical photogrammetric camera calibration. In comparison to the refractive index, there are no mathematical models that can describe these parameters, thus not permitting the achievement of a reliable camera calibration procedure. Eventually, it would be desirable, given an unrestricted underwater time, if a camera calibration could be completed in identical conditions to each archaeological photogrammetric dive.

3.2. Photogrammetric bundle adjustment

A matter of high importance is to have the acquired images prerectified before using them in the photogrammetric processing [11]. Preprocessing adjustments include radiometric enhancement of images, i.e. brightness adjustment, contrast enhancement, edge enhancement, noise reduction etc. Therefore, 46 images were selected and processed using Adobe Photoshop software, through 'Neutralize', 'Brightness-Contrast' and 'Color balance' commands (Figures 4a and 4b).

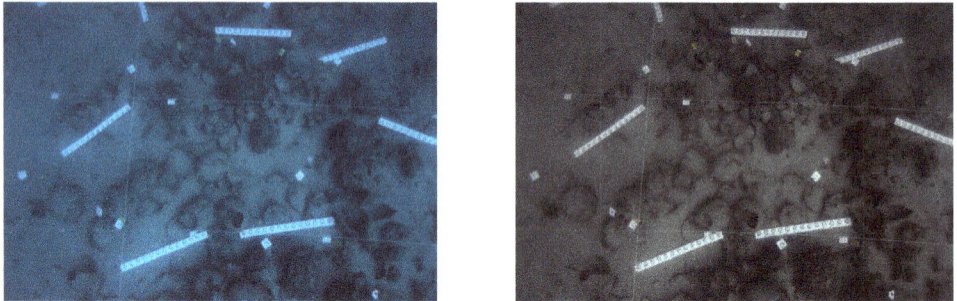

Figure 4: (a) Initial image, (b) Color processed image

Once all 46 images were preprocessed, the photogrammetric adjustment was implemented using Topcon's ImageStation® software. The block adjustment was done using 46 images at an approximate range of scale from 1:200 to 1:600, as the diver-photographer could not swim parallel to the site. Twenty (20) points with known coordinates, as resulted from the trilateration adjustment method with Site Recorder, were recognized, marked on the images and utilized as control points. The a priori control point precision was set at $\sigma_{XY} = 0.04$m and $\sigma_Z = 0.07$m. Table 1 shows the results of the bundle adjustment.

Table 1: Bundle adjustment results

RMSxy (μm)	6.2
Pixel size (μm)	5.8
RMSXY (m)	0.025
RMSZ (m)	0.037

A fact which has to be stressed is that the block included images of regions of sandy seabed. This was a problem, as far as finding a sufficient number of common points between images, is concerned. To increase the number of common points, Plexiglas strips with coded targets, similar to Photomodeler® calibration targets, were positioned in those sandy areas. This method proved to be very helpful in the end, in the attempt to evaluate the accuracy of the final orthophotomosaic of the site, especially in sandy areas where the control point distribution could not be very dense.

3.3. Orthophotomosaic of the shipwreck

One of the main goals of the work was to produce an accurate and radiometrically correct orthophotomosaic of the entire area of the shipwreck. Therefore, once the photogrammetric bundle adjustment of the block was completed, the extraction of a Digital Surface Model

Figure 5: Digital Surface Model of the site

was initiated. Approximately 54000 points and 2000 break lines were collected manually through stereoscopic vision of the oriented models, thus producing a dense DSM (Figure 5). The attempt for automated DSM extraction failed, due mainly to problems connected with a weak image radiometry, e.g. similar tones especially due to the absorption of the red wavelength, repetitive features and poor textures like sandy areas, scale variations and large rotation differences between images and, finally, occlusions. As a result, the acquired data did not reach the classical image matching methods standards and a manual DSM extraction was the only solution.

Once the DSM was extracted, an orthophotomosaic of the shipwreck's area, made out of 46 orthorectified images (Figure 6), was produced using Z/I´s ImageStation OrthoPro® software. The final product was evaluated in two ways:

1. By comparing the control points network, as it resulted from the trilateration adjustment, with the orthophotomosaic

2. By measuring on the orthophotomosaic the known distances between the Plexiglas strips of Photomodeler coded targets.

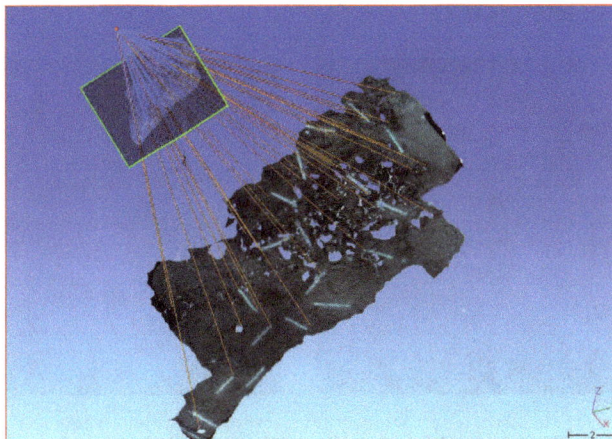

Figure 6: Orthophoto mapping of the shipwreck's area with the use of DSM

Figure 7: (a) 2D scaled drawing of a Brindisi amphora, (b) 3D revolved amphorae model, (c) 3D rendered model

3.4. 3D reconstruction of the shipwreck

The ship's cargo consisted mainly of amphorae of one type, i.e. Brindisi amphorae. A 3D theoretical revolved model was obtained using 2D scaled drawing of one such amphora, which was taken off the site using the Rhinoceros® software (Figures 7a, 7b and 7c).

Due to the fact that most of the objects were partially visible on images or even broken, photogrammetric measurements were not enough for the complete 3D reconstruction of the site. The main reason for not using the already constructed and dense DSM for plotting each object is that it is a 2.5 D plan. This means that many objects' attributes were not visible on images, thus hiding an important amount of information. Therefore, characteristic features of each object were photogrammetrically measured, so that accurate shape, size and direction of the amphorae could be restored. The final 3D model is a combination of the theoretical models and the photogrammetrically measured attributes of the various finds (Figure 8). The choice of attributes is based on measuring particular parts of the finds, i.e. rims, bodies, mouths or handles of each amphora, so that each object could be positioned as well as oriented efficiently. The photogrammetric measuring of those attributes was performed using Photomodeler® software, in which approximately 60 images were oriented. Photomodeler provides the opportunity of orienting a large number of images taken from various angles, thus regaining the lost information of hidden objects that led to the optimal 3D reconstruction of the site. Measured objects were divided into three layers in Photomodeler; a) the 'terrain' layer, which plays the role of a DTM and consists of points of sand and rocks, b) the 'finds' layer, which includes all measured attributes of finds and c) the 'control points' layer, which includes the control points network.

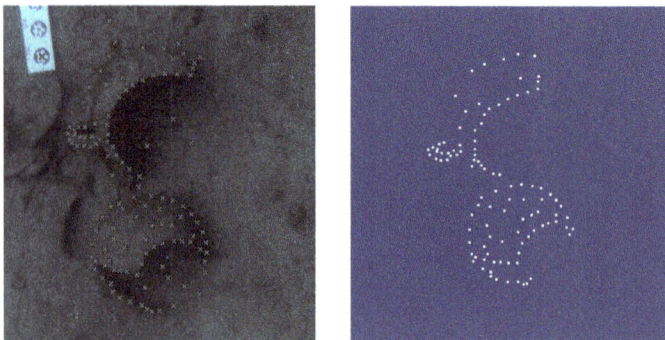

Figure 8: Measured attributes of the artifacts

A DXF file of points of the above layers, followed by images with the IDs of all points of interest upon (Figure 8), was extracted from Photomodeler® and then imported in Rhinoceros®. The entire 3D reconstruction of the site was implemented finally in Rhinoceros® by creating the terrain surface at first and placing, afterwards, each find in its right position (Figure 9).

A more realistic representation of the site was achieved by assigning texture extracted from images through a suitable rendering procedure (Figure 10). The rendering was implemented in Autodesk's 3DS max®.

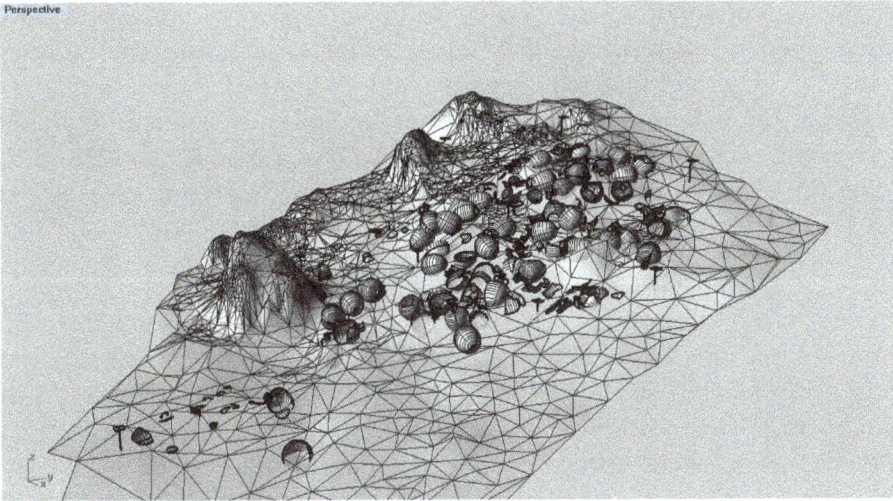

Figure 9: 3D wireframe model of the shipwreck

Figure 10: 3D rendered model of the shipwreck, Autodesk 3DS max

4. Conclusions

Estimating the accuracy of any underwater surveying technique is notoriously difficult [12]. The results of recording the shipwreck of the South Euboean gulf represent an attempt for the best possible achievable accuracy, when combining conventional tape measurements with modern and user-friendly photogrammetric software. Given the nature of the control points (some robust, others more fragile), it is likely that the points themselves had an uncertainty in their position, which affected the accuracy of the photogrammetric bundle adjustment. Therefore, the task of obtaining a robust control points' network seems as challenging as the task of using photogrammetry for underwater surveying without establishing a control points' network at all. The second task requires a perfectly calibrated camera and a way to restore the scale during the bundle adjustment of a block of images. Evaluating, finally, the work that has been done underwater, given the aforementioned available diving time, it seems that it should be preferable to consume more diving time on calibrating cameras under several conditions underwater, in order to obtain an optimal interior orientation and avoid to spend time on the really time consuming task of measuring distances with common tapes.

Moreover, the implementation of different software for one final goal may be over consuming too, in terms of time and work, but in this case, it was unavoidable to use several pieces of photogrammetric software. Each one was used as a different tool. Photomodeler, provides, firstly, the opportunity of a user-friendly automated camera calibration module and secondly, the opportunity of orienting a large amount of images taken from different angles, so that more information of the object could be obtained. On the other hand, ImageStation software was chosen as a more reliable way, comparing to Photomodeler®, to create a DSM and an orthophoto, thanks to the DSM collection through stereo vision.

In conclusion, the application of photogrammetry in terms of generating accurate and radiometrically efficient orthophotomosaics and 3D rendered models, combining inescapable traditional surveying techniques with contemporary digital software support, has proven to be a unique way to achieve a virtual exploration, a "digital trip", to an ancient shipwreck in deep waters, to a deep ancient cultural heritage.

References

[1] Gili Telem, Sagi Filin, Photogrammetric modeling of underwater environments, ISPRS Journal of Photogrammetry and Remote Sensing, 2010.

[2] Scarlatos D., Agapiou A., Rova M., Photogrammetric support on an underwater archaeological excavation site: The Mazotos shipwreck case, Cyprus, 2010.

[3] Canciani, M., Gambogi, P., Romano, G., Cannata, G., and Drap, P., 2002, Low cost digital photogrammetry for underwater archaeological site survey and artefact insertion. The case study of the Dolia Wreck in Secche della Meloria, Livorno, Italia, International Archives of Photogrammetry, Remote Sensing and Spatial Information Sciences 34.5/W12, 95–100.

[4] Holt, P., 2003, An assessment of quality in underwater archaeological surveys using tape measurements, IJNA 32.2, 246–51.

[5] Patias, P., 2006, Cultural Heritage Documentation. International Summer School 'Digital recording and 3D Modeling', Aghios Nikolaos, Crete, Greece, 24–29 April, www.photogrammetry.ethz.ch/summerschool/pdf/15_2_Patias_CHD.pdf, last updated 17 April 2006, accessed 27 July 2009.

[6] Drap P., Durand A., Provin R., Long L., Integration of multi-source spatial information and XML information system in underwater archaeology, Torino, 2005.

[7] Demesticha, S., The 4th-Century-BC Mazotos Shipwreck, Cyprus: A preliminary report, The International Journal of Nautical Archaeology, in press, 2010.

[8] Maas H., New developments in Multimedia Photogrammetry, Institute of Geodesy and Photogrammetry, Swiss Federal Institute of Technology, Zurich, 2000.

[9] Lavest J.M., Rivers G., and Lapreste J.T., Dry camera calibration for underwater applications, Machine Vision and Applications 13, pp. 245-253, 2003.

[10] Walford A., Personal communication Eos Systems Inc, Vancouver, 1996.

[11] Li R., Li H., Zou W., Smith R.G., and Curran T.A., Quantitative photogrammetric analysis of digital underwater video imagery, IEEE Journal of Oceanic Engineering, 22(2) : 364-375, 1997.

[12] Green J., Matthews S., Turanli T., Underwater archaeological surveying using PhotoModeler, VirtualMapper: different applications for different problems, The Nautical Archaeology Society, 2002.

## 4

Monitoring of a concrete roof using terrestrial laser scanning

Ján Erdélyi, Alojz Kopáčik, Ľubica Ilkovičová, Imrich Lipták, Pavol Kajánek

Slovak University of Technology, Faculty of Civil Engineering,
Radlinskeho 11, 813 68 Bratislava, Slovakia

Abstract

The paper deals with the geodetic monitoring of a parabolic shaped reinforced concrete roof structure in the chemical company Duslo, Ltd. in Šaľa (Slovak Republic). The monitored structure is a part of the roof of a warehouse used for the storage of fertilizer. The atmospheric conditions and the operation load caused deformation of the construction. For measurement was used the technology of terrestrial laser scanning. The displacements of the observed parts of the construction were calculated using planar surfaces by the procedure of Singular Value Decomposition of matrixes. The procedure of initial and 2 epochal measurements of deformations, the procedure of the data processing, and the results of the deformation monitoring are described.

Keywords: deformation monitoring, terrestrial laser scanning, reinforced-concrete construction, singular value decomposition

1. Introduction

The weather conditions and the operation load are causing changes in the spatial position and in the shape of engineering constructions, which affects their static and dynamic function and reliability. Because these facts, geodetic measurements are integral parts of engineering structures diagnosis.

This paper presents the geodetic monitoring of a parabolic shaped reinforced concrete roof construction of a fertilizer warehouse in the Duslo, Ltd., Šaľa, which is the largest chemical company in the Slovak republic. The operation load and the weather conditions caused shift between the blocks of the roof during the decades of operation. The measurements were done in 3 epochs during 2 months. The stability of the foundation strips of the construction was monitored by precise levelling. The deformation of the roof construction was measured using terrestrial laser scanning (TLS).

The advantage of TLS over conventional surveying methods is the efficiency of spatial data acquisition. TLS allows contactless determining the spatial coordinates of points lying on the surface on the measured object. The scan rate of current scanners (up to 1 million of points/s) allows significant reduction of time, necessary for the measurement; respectively increase the quantity of obtained information about the measured object. To increase the accuracy of results, chosen parts of the monitored construction can be approximated by single geometric entities using regression. In this case the position of measured point is calculated from tens or hundreds of scanned points (Vosselman et al., 2010).

2. Characteristics of the measured object

The measured object is used for storage of fertilizer in the chemical company Duslo, Ltd., Šaľa. It consists of a reinforced concrete construction with dimensions 30 m x 170 m and with height 14 m. On the roof in the middle part of the construction is situated a conveyor along the whole warehouse. The warehouse is founded on foundation strips (with dimensions 3.7 m x 172.0 m x 1.5 m) and is divided into 5 blocks.

Figure 1: The construction of the warehouse

The roof consists of a parabolic shaped reinforced concrete construction with parabolic transverse beams (with axial distance 4.8 m). The warehouse was built in year 1960. The operation load of the conveyor, and the weather conditions caused deformation of the roof construction during the decades of operation. The mentioned reasons caused a shift approximately 150 mm between the 1st and the 2nd block, which is visible at the dilatation.

The aim of the measurements was the geodetic monitoring of the parts of the roof construction near the dilatation joints, and the determination of the rate of changes.

3. Deformation monitoring

Considering the unclear cause of deformations, the monitoring was designed to be able not only to quantify the movements of the mentioned parts of the roof structure, but even the eventual motions of foundations. The measurements were done in 3 epochs during 2 months, on October 7th, October 21st and December 2nd 2013. The aim of the monitoring was to determine the rate of the displacements, and to determine their influence to the secure operation. The deformations of the roof construction were monitored using terrestrial laser scanning, and the behavior of the foundations was measured by precise levelling.

3.1. Precise levelling

The monitoring of the foundation strips was performed in 3 measurement epochs (mentioned above) using precise levelling. The height of 8 measured points (N1.1-N2.4) was determined relative to the height of 3 control points (VB1-VB3) in a local height system (Fig. 2). The control points are situated near the monitored object, in the footings of the pylons of a pipeline nearby the warehouse. These are stabilized by ground benchmarks. The stability of the reference net was determined comparing the height differences between the points in each epoch.

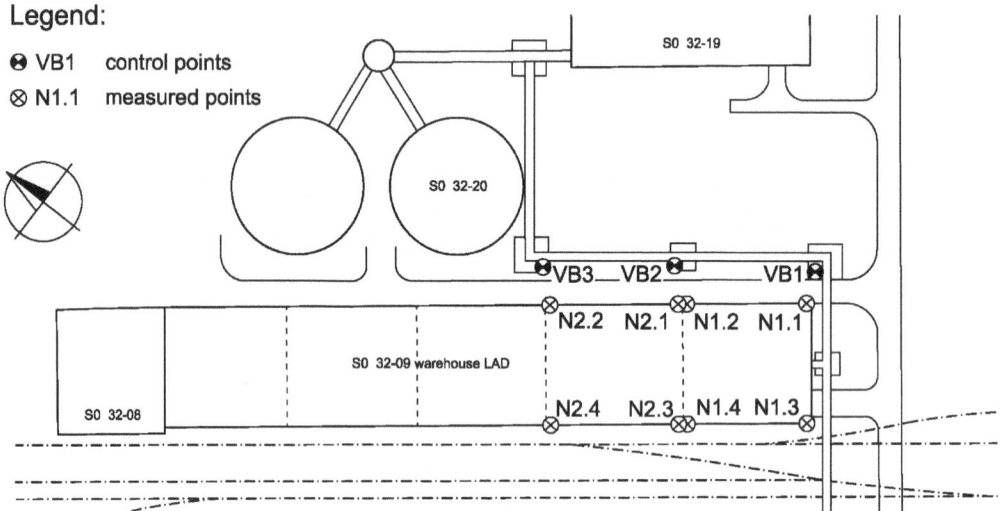

Figure 2: Position of measured and control points – precise levelling

The measured points are situated on the beginning and end of 1st and 2nd block on the both sides of the warehouse. The points are stabilized by wall benchmarks in the bottom part of the parabolic transverse beams. The vertical displacements of these points were determined as the difference between the heights of these points in each epoch.

Table 1: Vertical Displacements – Precise Levelling

| Point No. | Height of points | Vertical displacements of observed points considering to initial measurement (October 7th 2013) | | | | | |
| | | October 21th 2013 | | | December 2nd 2013 | | |
	H [mm]	h [mm]	σ_h [mm]	Decision	h [mm]	σ_h [mm]	Decision
N1.1	100.2167	0.0	0.28	no shift	+0.7	0.28	5%
N1.2	100.0306	-0.3	0.28	5% - 30%	-0.6	0.28	5% - 30%
N1.3	99.7724	+0.1	0.28	no shift	----	--	----
N1.4	100.2479	+0.1	0.28	no shift	-0.4	0.28	5% - 30%
N2.1	99.9793	-0.1	0.28	no shift	+0.6	0.28	5% - 30%
N2.2	100.0296	-0.6	0.28	5% - 30%	-0.2	0.28	no shift
N2.3	100.2393	+0.1	0.28	no shift	-0.3	0.28	5% - 30%
N2.4	99.9995	-0.5	0.28	5% - 30%	0.0	0.28	no shift

The statistical significance of the displacements was determined on the basis of the statistical analysis using interval estimates. The measurement did not shown any displacements on most of the observed points; respectively, the risk of the decision is 5% - 30%. Due to the results of precise levelling, it can be assumed that the foundation strips of the structure are stable, or the movements are slow without influence to the security.

3.2. Terrestrial laser scanning

The monitoring of the roof structure was performed using TLS Leica ScanStation2. The bottom side of the roof was scanned from a single position of the scanner. Scanned was a 1 m wide strip on the both sides of the dilatation (Fig. 3). It was not possible to scan strip along the whole dilatation, because the fertilizer was not removed from the left side of the mentioned part of the structure. The scanner was positioned in each epoch approximately in the same position to ensure the same conditions of the measurements (distance from the scanner, angle of incidence of the measuring signal).

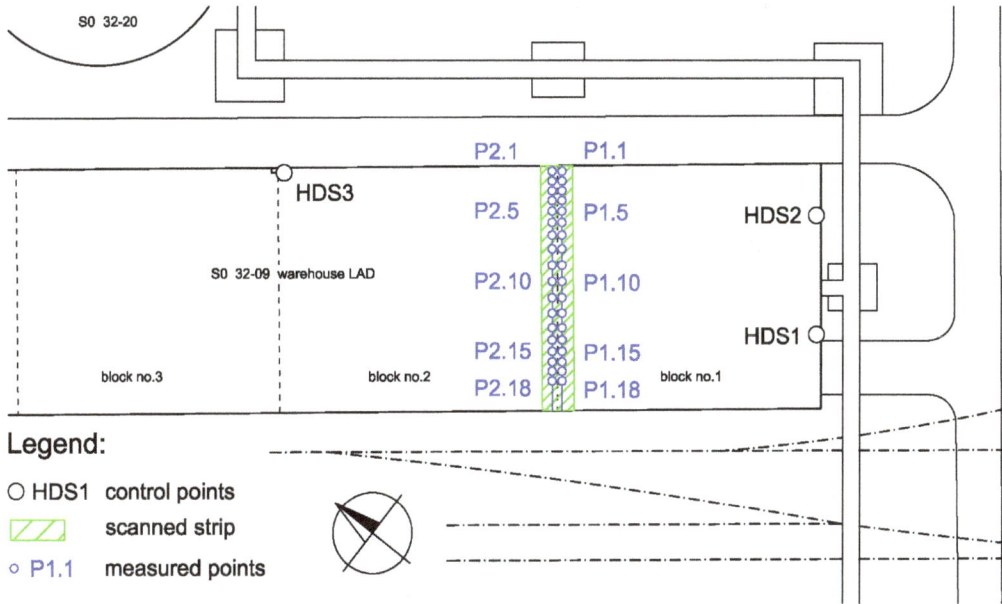

Figure 3: Position of measured and control points – terrestrial laser scanning

The reference network consists of three control points stabilized on the pillars of the warehouse frame by metallic fasteners (it was possible due to the stability of the foundations). All of the control points were signalized by the Leica HDS targets. To improve the efficiency of the measurements, a simple script was defined before scanning in each epoch. This script defines a separate field of scanning for different parts of the construction, the scan resolution in each field, and the target acquisition. The minimal point density on the surface of the roof was 10 mm x 10 mm.

The data obtained by the TLS were transformed to a local coordinate system. The accuracy of the transformation was calculated from the differences between the common identical points after the transformation. The main task of the data processing was modelling the position of the measured points by small planar surfaces. These are positioned on the bottom side of the roof every 2 m on the both sides of the dilatation (Fig. 3).

The vertical displacements of the measured points were determined as the difference between the heights of these points in each epoch. The height of the points was calculated using orthogonal regression. The vertical displacements were recalculated to orthogonal displacements along the normal vector to the surface in each part of the structure. During the data processing of the initial measurement, square fences of 0.1 m x 0.1 m were defined on the bottom side of the roof. These fences approximately define the same set of points in each epoch.

Orthogonal regression is calculated from the general equation of a plane by applying Singular Value Decomposition:

$$\mathbf{A} = \mathbf{U\Sigma V}^{\mathrm{T}}$$

where $\mathbf{A}$ is the design matrix, with dimensions $n \times 3$, and n is the number of points used for the calculation. The design matrix contains the coordinates of the point cloud reduced to a centroid. The column vectors of $\mathbf{U}^{nxn}$ are normalized eigenvectors of matrix $\mathbf{AA}^{\mathrm{T}}$. The column vectors of $\mathbf{V}^{3x3}$ are normalized eigenvectors of $\mathbf{A}^{\mathrm{T}}\mathbf{A}$. The matrix $\mathbf{\Sigma}^{n \times 3}$ contains eigenvalues on the diagonals. Then the normal vector of regression plane is the column vector of $\mathbf{V}$ corresponding to the smallest eigenvalue from $\mathbf{\Sigma}$ (Lacko, 2008, Čepek, 2009).

The mean errors of the displacements were obtained using the law of propagation of uncertainty, from the mean error of the transformation and the mean error of the regression planes, which were calculated from the orthogonal distance of points from these planes. The statistical significance of the displacements was determined on the basis of the statistical analysis using interval estimates (Kopáčik et al., 2013). Table 2 shows the displacements (vertical and orthogonal) of the selected points observed.

Table 2: Displacements – Terrestrial Laser Scanning

Point No.	Displacements of observed points considering to initial measurement (October 7th 2013)									
	October 21th 2013					December 2nd 2013				
	Δz [mm]	σ [mm]	Decision	Orth. disp. [mm]	Decision	Δz [mm]	σ [mm]	Decision	Orth. disp. [mm]	Decision
P1.1	0	1.7	no shift	0	no shift	0	1.4	no shift	0	no shift
P1.2	0	2.2	no shift	0	no shift	-3	1.6	5%- 30%	-1	no shift
P1.3	-2	2.1	no shift	-1	no shift	1	1.7	no shift	1	no shift
P1.4	-2	1.8	no shift	-1	no shift	1	1.5	no shift	1	no shift
P1.5	-3	1.6	5%-30%	-2	5%- 30%	-1	1.5	no shift	-1	no shift
P1.6	-4	1.4	5%	-2	5%- 30%	-2	1.3	5%- 30%	-2	no shift
P1.7	-4	1.4	5%	-3	5%- 30%	-4	1.1	5%	-3	5%- 30%
P1.8	-3	1.4	5%	-2	5%- 30%	-5	1.3	5%	-4	5%
P1.9	-3	1.7	5%-30%	-3	5%- 30%	-5	1.3	5%	-5	5%
P1.10	-1	2.2	no shift	-1	no shift	-3	1.9	5%- 30%	-3	5%- 30%
P1.11	0	2.2	no shift	0	no shift	-4	1.6	5%	-4	5%
P1.12	0	2.1	no shift	0	no shift	-4	1.6	5%	-3	5%- 30%
P1.13	2	2.2	no shift	2	no shift	-1	1.7	no shift	-1	no shift
P1.14	2	2.1	no shift	2	no shift	0	1.9	no shift	0	no shift
P1.15	3	2.1	5%-30%	2	no shift	2	1.9	no shift	1	no shift
P1.16	4	2.0	5%-30%	2	no shift	3	1.6	5%- 30%	2	5%- 30%
P1.17	2	1.8	no shift	1	no shift	0	1.5	no shift	0	no shift
P1.18	0	1.9	no shift	0	no shift	-1	1.8	no shift	0	no shift

The measurement did not show any displacements on most of the observed points; respectively, the movements are slow with the risk of the decision 5% - 30%. The Fig. 4 shows the graphical representation of the displacements of the selected points between the initial measurement (October 7th) and the 2nd measurement epoch (October 21st).

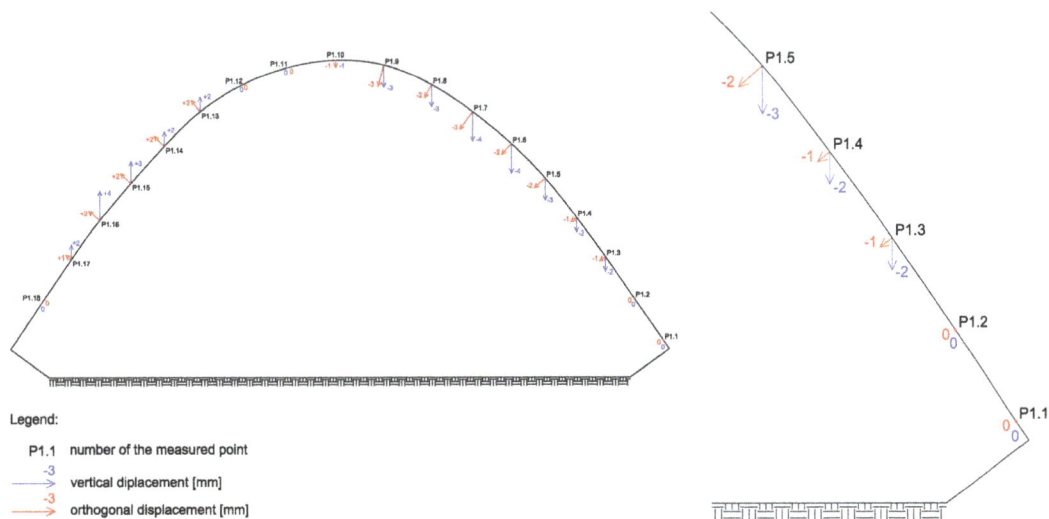

Figure 4: Displacements in the 1st control epoch (right), and the detail of vertical and orthogonal displacements (left)

4. Conclusion

The aim of the monitoring described in this paper was to determine the rate of the displacements, and to determine their influence to the secure operation of the warehouse. The results did not show significant displacement of the monitored part of the warehouse construction (see Table 1 and Table 2). Due to these facts the measurements planned for the future were cancelled. Mechanical measuring equipment was mounted on the mentioned part of the structure. The eventual deformation will be monitored visually, by the operating staff of the warehouse.

References

[1] Kopačik, A. et al. 2013. Deformation Monitoring of Bridge Structures Using TLS. In *2nd Joint International Symposium on Deformation Monitoring* [USB]. Nottingham: University of Nottingham, 2013, 8 p.

[2] Lacko, V. 2008. *Singular Value Decomposition and Difficulties of Software Implementation of Golub Algorithm and its Determination*: Student sciense conference. Bratislava: Comenius University in Bratislava, 2008. 69 p.

[3] Vosselman, G. – Maas, H. G. 2010. *Airborn and Terrestrial Laser Scanning*. Dunbeath: Whittles Publishing, 2010. 318 p. ISBN 978-1904445-87-6.

[4] Čepek, A. and Pytel, J. 2009. A note on numerical solutions of least squares adjustment in GNU project gama, In Pilz J., editor, *Interfacing Geostatistics and GIS*, Springer Berlin Heidelberg, pp. 173-187. doi:10.1007/978-3-540-33236-7_14

5

Quality parameters of digital aerial survey and airborne laser scanning covering the entire area of the Czech Republic

Jiří Šíma
Novorossijská 18, Praha 10, Czech Republic
jirka.sima@quick.cz

Abstract

The paper illustrates the development of digital aerial survey and digital elevation models covering the entire area of the Czech Republic at the beginning of 21st century. It also presents some results of systematic investigation of their quality parameters reached by the author in cooperation with Department of Geomatics at the Faculty of Applied Sciences of the University of Western Bohemia in Pilsen and the Land Survey Office.

Keywords: digital aerial survey, orthophoto imagery, aerial laser scanning, digital elevation model, digital surface model, Czech Republic

1. Introduction

Year 2010 became a turning point as far as gathering of up-to-date geospatial data from the whole territory of the Czech Republic concerned. Aerial survey with digital cameras in RGB and NIR spectral bands with on-the-ground resolution of 0.20 m covered 100 % of the state territory in the course of 2010-12. Periodically repeated digital image records serve as a database for photogrammetric processing – preferentially of digital colour orthoimagery of the Czech Republic with on-the-ground resolution of 0.25 m. The era of taking aerial photographs on film and their subsequent rastering by means of precise photogrammetric scanners was de facto closed for this purpose. At the same time aerial laser scanning of 68.4 % of the state territory was successfully realized within the Project of New Hypsometry of the Czech Republic [1]. Its main goal is to provide bodies of state administration with High Resolution Elevation Data in form of Digital Elevation Model and Digital Surface Model of the entire state territory by 2015. New models of hypsometry will fulfil all requirements of the INSPIRE project as well. The Czech Office for Surveying, Mapping and Cadastre, Ministry of Defence and Ministry of Agriculture are investors of both projects. Land Survey Office and Military Geographical and Hydrometeorological Office are main compilers of digital image and LIDAR data. The author and Department of Geomatics at the Faculty of Applied Sciences of the University of West Bohemia in Pilsen have collaborated in definition and evaluation of quality parameters of all resulting products from the very beginning of development of all above mentioned projects.

2. Development of aerial survey for production of the CZ Orthophoto

In order to reach complete coverage of the Czech Republic with aerial photographs or digital images in 3year period total area of 78 865 km2 was divided into 3 zones (West, Central, East) in 2003 respecting the layout of the State map series in scale 1 : 5000 (Fig.1). From 2003 to

2008 wide angle aerial photographs, mostly in scale 1 : 23 000 from relative flight height 3500 m have been exposed on colour negative film and transformed into raster form by precise photogrammetric scanners. Such parameters allowed to produce digital colour orthophoto imagery with pixel size 0.50 m on the ground. Since 2009 orthophoto imagery with higher resolution (0.25 m on the ground) has been required and other survey flight parameters - photo scale 1 : 18 000 and lower flight height 2740 m have to be chosen.

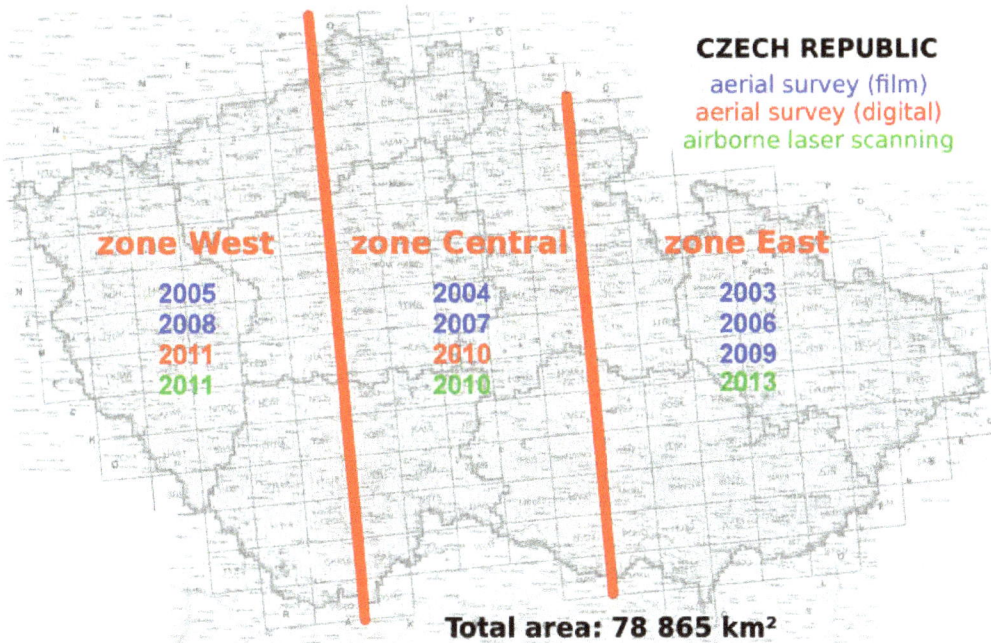

Figure 1: Schedule of aerial survey and airborne laser scanning of the Czech Republic

Since 2010 aerial photography on a film has been replaced by digital aerial survey using digital large format metric cameras for simultaneous registration of images in PANCHRO, R,G,B and NIR spectral bands. Minimum size of a sensor element 6 micrometers allowed to reach economically the on-the-ground resolution between 20 and 25 cm of the pixel size. All private firms taking part in tenders for digital aerial survey have used Vexcel UltraCam X or Xp cameras. Table 1 shows a complete review of methods used for aerial survey of the entire area of the Czech Republic [4].

Digital aerial survey covered the Central zone in 2010 and the zone-West in 2011. The rest of state territory (22.8 % - almost 28 thousand km^2 in zone-East) was imaged in 2012 commonly with one half of the Central zone, because there is an intention to reduce recent 3-year period into 2-year period since 2012. (Fig. 2). All projects of aerial surveys have been funding by the Czech Office for Surveying, Mapping and Cadastre and the Ministry of Agriculture.

The main product of periodical aerial surveys – the CZ Orthophoto (Ortofoto ČR in Czech) serves first of all for needs of state authorities and public administration. Some important examples of its application should be introduced here:

- updating of the Land Parcel Information System; orthophoto is documenting the areas recently cultivated by farmers when they ask for financial support from the funds of

Year	Method of aerial survey and rastering	Average scale of photographs / images	On-the ground pixel size (pixel of CZ Orthophoto)
2003-2008	analogue, on colour film + scanning into the raster form	1 : 23 000	0.46 – 0.48 m (0.50 m)
2009	analogue, on colour film + scanning into the raster form	1 : 18 000	0.27 m (0.25 m)
2010	digital (PAN, R, G, B, NIR) direct raster registering – UC XP	1 : 32 000	0.19 m (0.25 m)
2011	digital (PAN, R, G, B, NIR) direct raster registering – UC X, XP	1 : 35 000	0.21 – 0.25 m (0.25 m)
2012	digital (PAN, R, G, B, NIR) direct raster registering – UC XP	1 : 36 000	0.22 m (0.25 m)

Table 1: Recent development of aerial survey parameters for the CZ Orthophoto production

European Union,

- updating of the Fundamental Base of Geographic Data (known under acronym ZABAGED ®) that is a topological-vectorial spatial data base at level of 1 : 10 000 Base Map of the Czech Republic,

- updating of a military Digital Landscape Model 25, a similar database at level of 1 : 25 000 Military Topographic Map,

- providing the Infrastructure for Spatial Information in Europe (INSPIRE) with recent orthophoto imagery of the Czech Republic.

- as an substantial part of the Digital Map of Public Administration.

Figure 2: Digital aerial survey of the Czech Republic in two-year interval since 2012

3. Quality parameters of the CZ Orthophoto

The quality of the CZ Orthophoto enables to use this product to some other tasks [6]:

- discovering some discrepancies, gross and systematic positional errors in planimetric representation of objects in digitized cadastral maps (see Fig. 3),

- colour infrared orthophotos are widely used to national inventory of forests.

From 2004 the Laboratory of digital photogrammetry at the Faculty of Applied Sciences in Pilsen, and later the author in collaboration with Department of land surveying of the Land Survey Office have been continuously assessing the absolute positional accuracy of the CZ Orthophoto [4] using more test fields containing hundreds of check points measured mostly by GPS Real Time Kinematic method with positional accuracy better than 10 cm, and identical points well identified in the orthophoto. Remarkable seems to be high absolute positional accuracy of orthophoto from digital images in spite of their distinctly lesser scale [5] [8] [9] (see Table 2).

Method of image data registering	Number of points	cY [m]	cX [m]	mY [m]	mX [m]	mXY [m]	ΔYmax [m]	ΔXmax [m]
colour film 1 : 16 650 (2008) raster scanning pixel size 20 μm	732	-0.17	0.08	0.36	0.33	**0.35**	1.88	- 1.67
digital images 1 : 32 000 (2010) pixel 6 or 7.2 μm	430	0.05	-0.03	0.14	0.16	**0.15**	0.55	0.60
digital images 1 : 35 000 (2011) pixel size 6 μm	301	0.02	0.07	0.21	0.24	**0.23**	0.63	0.89
digital images 1 : 36 000 (2012) pixel size 6 μm	90	.0,04	-0.05	0.20	0.23	**0.22**	0.48	0.53

Explanations: c ...   systematic error
  m   root mean square error
  Δmax maximum error

Table 2: Results of testing the absolute positional accuracy of the CZ Orthophoto

Following check point types have been chosen for evaluation of absolute positional accuracy of the CZ Orthophoto:

- visible on-the-ground corners of a building

- foot of a pylon, telegraph pole or lamppost

- on-the-ground corner of a fence, underpinning or masonry wall

- midpoint of a circular manhole, drainage and well cover,

- corner of a kerb or road surface

- X or T form of white line intersections on a road surface

Figure 3: Original cadastral survey of a local road 50 years ago and reality as shown in the CZ Orthophoto

High absolute positional accuracy of the CZ Orthophoto made from digital aerial images in 2010-12 (see Tab. 3) has been reached thanks to compliance with keeping of three principles:

1. targeting of optimally distributed ground control points within the block of aerial images; predominantly of monumented triangulation points having known and accurate absolute position defined in the compound coordinate reference system (S-JTSK, Balt-po vyrovnání). Their number should not be less than 1 per 25 digital images, e.g. 40 GCP within the block of 1000 images,

2. on-the-board registration of elements of exterior orientation by GPS/IMU apparatus that makes the spatial geometry of a block stronger,

3. modern and effective software for digital automatic aerotriangulation (AAT) that enables quickly change the input parameters and repeat the computation (MATCH-AT version 5.2.1 or 5.4.2 has been used in two processing centres).

4. Development of digital elevation models in the Czech Republic

Until the year 2000 the Czech Republic was completely covered with hypsography based on graphical contour lines with two-metre interval represented in the Base Map of the Czech Republic in scale 1 : 10 000 (Fig. 4). In the course of establishing the Fundamental Base of Geographic Data (ZABAGED®) in topological-vectorial form, above mentioned hypsography has been digitized into a 3D model forming the Triagulated Irregular Network called

Year	Number of AAT blocks	RMSE of residuals on ground control points use for AAT			Number of AAT blocks	RMSE of residuals on check control points not use for AAT		
		m_X AAT [m]	m_Y AAT [m]	m_H AAT [m]		M_X [m]	M_Y [m]	M_H [m]
2010	17	0.113	0.100	0.200	12	0.160	0.140	0.300
2011	17	0.089	0.080	0.216	11	0.111	0.104	0.268
2012	23	0.067	0.074	0.137	22	0.184	0.173	0.256

Table 3: Results of testing the accuracy of digital automatic aerotriangulation

ZABAGED®-výškopis 3D vrstevnice. Another digital elevation model (DEM) in the form of 10 x 10 m grid called ZABAGED® - 10x10 m grid was derived from the previous one.Typical elevation accuracy of those models is: 0.7 - 1.5 m in open terrain without continuous vegetation cover, 1 - 2 m in built-up areas and 2 - 5 m in forests (accidentaly up to 20 m in the mountains).

Figure 4: Original contour lines in the Base Map 1 : 10 000 and their TIN representation after vectorizing

More demanding requirements on elevation accuracy for production of orthophotos with high on-the-ground resolution (25 cm and better) and frequent demands for 3D modelling of terrain ground and surface iniciated generation of the Project of New Hypsometry of the Czech Republic within the period from 2009 to 2015. As the most effective method the airborne laser scanning (ALS) was chosen for this purpose [1] [7].

Figure 5: Laser scanning system LITEMAPPER on the board of L-410 FG aircraft

In contrast with gathering the aerial photos and digital images by Czech or foreign private firms only, the airborne laser scanning has been accomplished by three state administration authorities: Ministry of Defence, Ministry of Agriculture and the Czech Office for Surveying, Mapping and Cadastre. Their tasks have been distributed according to disponible capacities.

Army of the Czech Republic dispones with a special photogrammetric aircraft L-410 FG of the Czech production suitable for both airborne laser scanning and digital aerial survey as well. Usual speed of a survey flight or airborne laser scanning is 250 km per hour. The flight hights range from 1200 m to 3200 m. In case of airborne laser scanning there is a Laser Scanning System LITEMAPPER hired out by German firm IGI installed on the board of L-410 FG (Fig.5). Parts of that system are: Laser Scanner RIEGL LMS Q680, GPS NovaTel equipment and IMU produced by the firm IGI.

Fig. 6 illustrates state of airborne laser scanning at the end of 2011. In 2012 the flights with L-410 FG had to be frozen through the necessity of its general overhaul. Individual blocks for laser scanning and data processing occupy the area from 10 x 10 km up to 10 x 40 km depending on maximum difference of elevations inside. The flight lines are parallel to E-cordinate axis of UTM projection used by the Army of the Czech Republic. There are two UTM 6-degree zones in the Czech Republic but a unique national coordinate reference system (S-JTSK) generally used by the civilian sector (e.g. in cadastre).

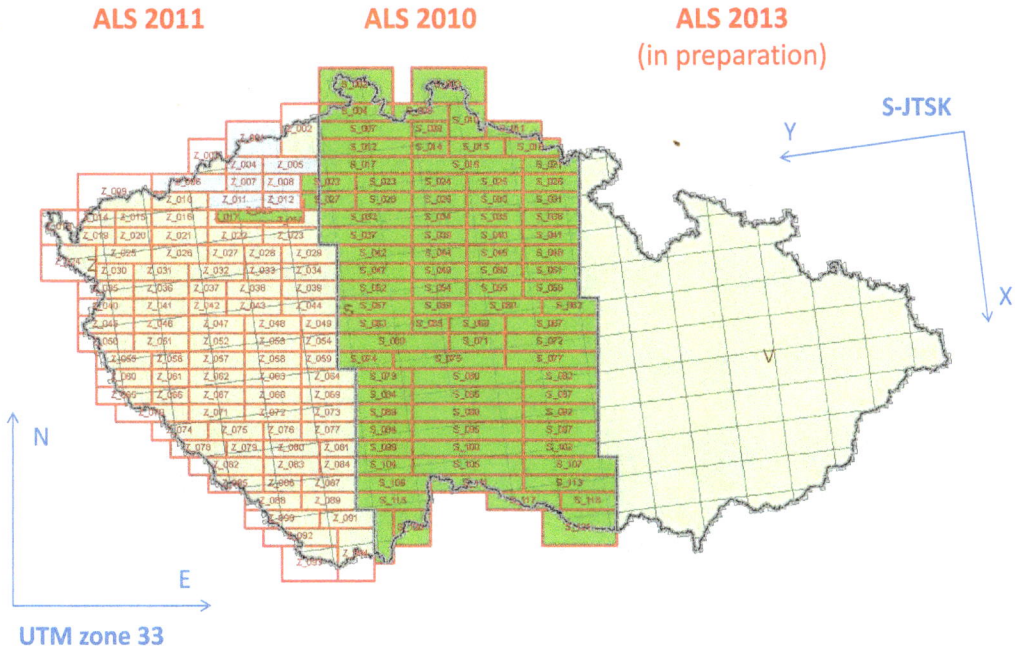

Figure 6: Coverage of the Czech Republic with airborne laser scanning data (March 2013)

5. Quality parameters of new digital elevation models

There are three final products of aerial laser scanning within the bounds of the Project of New Hypsomentry of the Czech Republic:

- Digital Elevation Model DMR 4G as a grid 5 x 5 m oriented paralelly to axes of national coordinate reference system S-JTSK or to the UTM grids in zones 33 and 34 (Fig. 7). This product should be always ready for distribution till 6 months after gathering ALS data. Its implicit elevation accuracy given by RMSE is 30 cm in open terrain and 1 metre in forested area. Suitable applications are: high resolution ortophoto production, draining of precipitation from a catchment basin, modelling of ecological disasters [2].

- Digital Ground Model DMR 5G in the form of Triangulated Irregular Network representing most of terrain break lines too should be ready for distribution till 12 months after gathering ALS data. Its implicit elevation accuracy given by RMSE is 18 cm in open terrain and 30 cm in forested area. Suitable applications are: new hypsometry of the Czech Republic (especially contour lines for State map series in scale from 1 : 5000 to 1 : 50 000), modelling of floods, modern spatial planning [3].

- Digital Surface Model (DMP 1G) ready for distribution till 18 months after gathering ALS data. Its implicit elevation accuracy given by RMSE is 40 cm on solid objects or bare ground and 70 cm on the surface with full-grown vegetation. Suitable applications are: optical visibility in rough terrain, night flying of helicopters, propagation of electromagnetic waves, true orthophoto in built-up areas.

The Land Survey Office organized a comprehensive testing for accuracy assessment of the DMR 4G using 240 horizontal test areas in Central zone of the Czech Republic (each equipped

Figure 7: DMR 4G covering 68,4 % of the Czech Republic (March 2013)

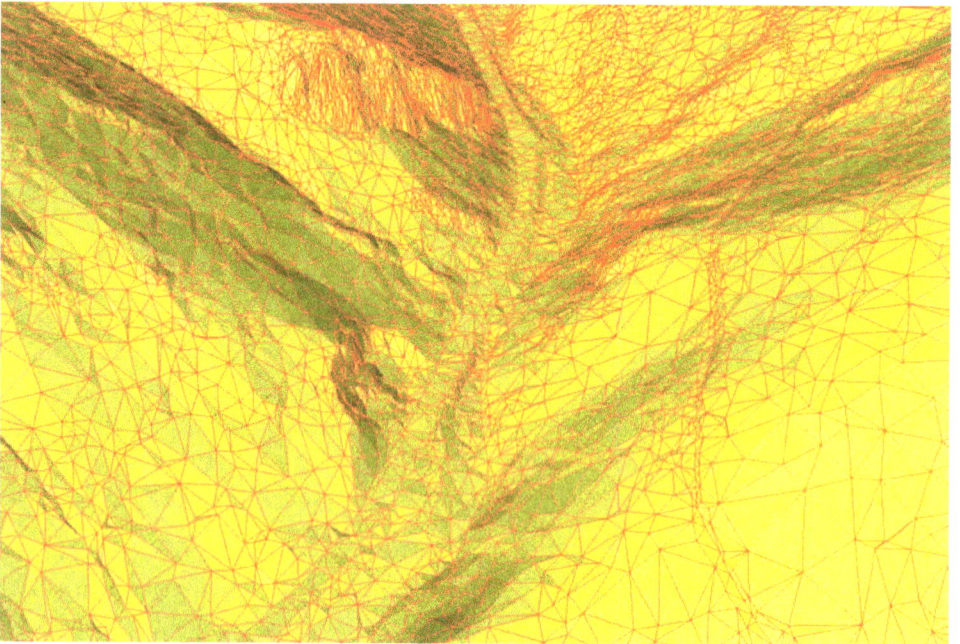

Figure 8: DMR 5G covering 33,4 % of the Czech Republic (March 2013)

with 30 – 100 check points) for determination of an average systematic error in elevations within the whole area mentioned above. It was –0.15 m (under the ground surface). After mass elimination of this systematic error the standard deviation reached 0.08 m only!

970 representative points of terrain surfaces were chosen and geodetically measured in 6 terrain types with various land cover. The results of comparison with identical points (Table 4) interpolated from the grid 5 x 5 m showed that expected elevation accuracy has been generally reached except for break lines of roads, embankments and trenches where this grid

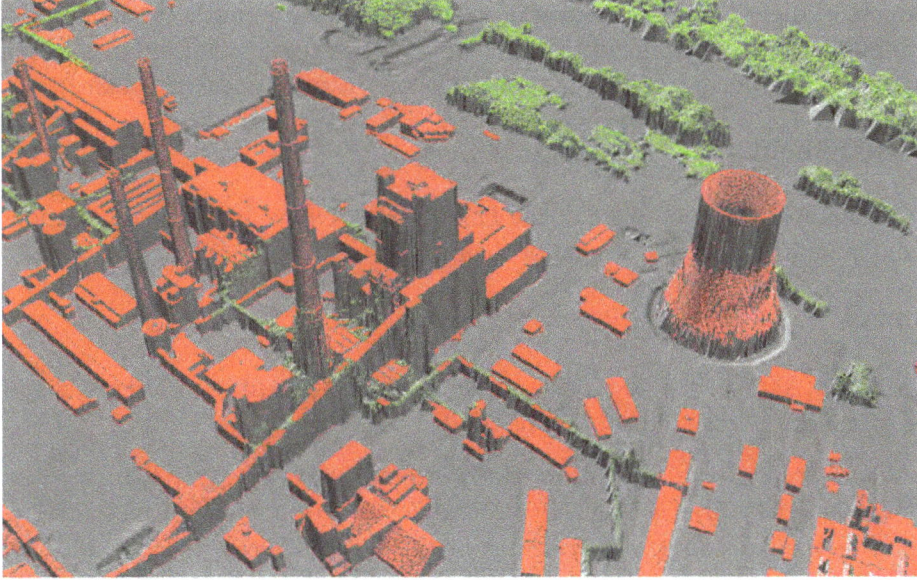

Figure 9: DMP 1G covering 32,9 % of the Czech Republic (March 2013)

is too smooth for such an application [2].

Category of surface and land cover	Systematic error [m]	RMSE (H) [m]	Maximum error [m]
Roads and highways	-0.25	0.34	0.77
Hard surfaces without vegetation	-0.01	0.07	0.26
Parks in built-up areas	-0.09	0.14	0.22
Arable land	-0.01	0.13	0.66
Meadows and pastures	-0.09	0.18	0.85
Shrubs , parkways, forests	-0.02	0.13	0.85
Average value	**-0.08**	**0.17**	**0.60**

Table 4: Results of testing the DMR 4G elevation accuracy

That's why the main product of aerial laser scanning of the whole state territory will be the Digital Ground Model of 5th generation (DMR 5G). Its density (up to two points per square metre) allows to represent most of important terrain break lines, but it will be appropriately reduced on plane or less curved surfaces [3].

Category of surface and land cover	Systematic error [m]	RMSE (H) [m]	Maximum error [m]
Break lines of roads and highways	-0.11	0.18	0.66
Hard surfaces without vegetation	-0.09	0.13	0.37
Arable land	-0.07	0.14	0.56
Meadows and pastures	-0.03	0.21	0.42
Shrubs, parkways, forests	-0.06	0.13	0.46
Average value	**-0 .07**	**0.14**	**0.49**

Table 5: Results of testing the DMR 5G elevation accuracy

All products of digital aerial survey and airborne laser scanning from the entire area of the Czech Republic have been already/or will be step-by-step at disposal on Geoportal of the Czech Office for Surveying, Mapping and Cadastre (http://geoportal.cuzk.cz).

Acknowledgment

The author appreciates employees of the Land Survey Office for their cooperation, data providing and submitting several results of their analyses.

References

[1] Brázdil, K. (2009): Projekt tvorby nového výškopisu území České republiky. Geodetický a kartografický obzor, 2009, č. 7, pp. 145-151.

[2] Brázdil, K. et al (2010): Technická zpráva k digitálnímu modelu reliéfu 4. generace (DMR 4G). Zeměměřický úřad a Vojenský geografický a hydrometeorologický úřad. Dostupné z http://geoportal.cuzk.cz/Dokumenty/TECHNICKA_ZPRAVA_DMR_4G_15012012.pdf

[3] Brázdil, K. et al (2011): Technická zpráva k digitálnímu modelu reliéfu 5. generace (DMR 5G). Zeměměřický úřad a Vojenský geografický a hydrometeorologický úřad. Dostupné z http://geoportal.cuzk.cz/Dokumenty/TECHNICKA_ZPRAVA_DMR_5G.pdf

[4] Brázdil, K. et al (2012): Technická zpráva k ortofotografickému zobrazení území ČR (Ortofoto České republiky). Zeměměřický úřad a Vojenský geografický a hydrometeorologický úřad. Dostupné z www.linkon.cz/gao6p

[5] Šíma, J. (2009): Průzkum absolutní polohové přesnosti ortofotografického zobrazení celého území České republiky s rozlišením 0,50, 0,25 resp. 0,20 m v území na Západočeské univerzitě v Plzni. Geodetický a kartografický obzor, 2009, č.9, pp. 214-220.

[6] Šíma, J. (2010): Nové zdroje geoprostorových dat pokrývajících celé území státu od roku 2010 – první výsledky výzkumu jejich kvalitativních parametrů. In: Sborník sympozia GIS Ostrava 2011. VŠB-TU Ostrava, 2011(nestránkováno). ISBN 978-80-248-2366-9.

[7] Šíma, J. (2011): Příspěvek k rozboru přesnosti digitálních modelů reliéfu odvozených z dat leteckého laserového skenování celého území ČR. Geodetický a kartografický obzor, 2011, č. 5., pp. 101-106.

[8] Šíma, J. (2013): Digitální letecké měřické snímkování – nový impulz k rozvoji fotogrammetrie v České republice. Geodetický a kartografický obzor, 2013, č.1, pp. 15-21. Dostupné z http://egako.eu/wp-content/uploads/2012/11/gako_2013_01.pdf

[9] Švec, Z. (2013): Absolutní polohová přesnost Ortofota ČR vyhotoveného z digitálních leteckých měřických snímků . Geodetický a kartografický obzor, 2013, č.2, pp.31-38. Dostupné z http://egako.eu/wp-content/uploads/2012/11/gako_2013_02.pdf

## 6

Development and testing of INSPIRE themes Addresses (AD) and Administrative Units (AU) managed by COSMC

Michal Med, Petr Souček

The Czech Office for Surveying, Mapping and Cadastre (COSMC)
Prague, the Czech Republic

`michal.med@cuzk.cz, petr.soucek@cuzk.cz`

Abstract

Main content of this article is to describe implementing INSPIRE themes Addresses and Administrative Units in Czech Republic. Themes were implemented by Czech Office for Surveying, Mapping and Cadastre. Implementation contains developing GML files with data and designing its structure, developing and testing of INSPIRE services and preparing metadata for data and services. Besides harmonised INSPIRE themes COSMC manages also non-harmonised themes Cadastral map (KM) and Units eXtended (UX).

Keywords: INSPIRE, Cadastre, Addresses, Cadastral Parcels, Administrative Units, Buildings, Metadata, RÚIAN, services, WMS, WFS, GML

1. Introduction

INSPIRE – **IN**frastructure for **SP**atial **InfoR**mation in **E**urope is a Directive of European Commission and Council, which was transposed into Czech legislation in 2009 by the law number 380/2009 Col., which amends laws number 123/1998 Col., on the right to information about environment and number 200/1994 Col., about surveying.

Figure 1: INSPIRE logo

From the law number 123/1998 Col. come (among others) following duties:

- create and manage metadata for spatial data files and for network services,

- harmonise spatial data sets according to the Directive,
- create interoperable network services.

Important part of implementation is also giving information on implementation to public. All basic information about implementation of data and services are available at geoportal COSMC[1] in czech and in english in a bookmark INSPIRE. Pages for themes Cadastral Parcels (CP), Addresses (AD) and Administrative Units (AU) have a special look and structure. It was designed for better intelligibility of data and services for users. From the geoportal, there is also possibility of downloading data and access network services. Data, services and informations are available on national INSPIRE geoportal[2], administered by CENIA, Czech environmental agency.

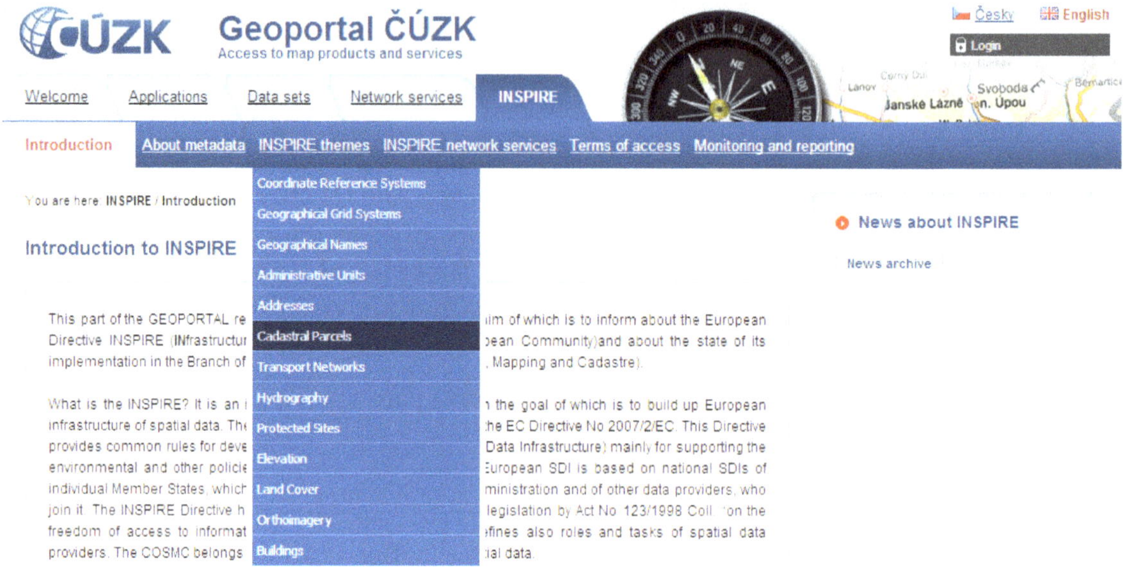

Figure 2: COSMC geoportal

This geoportal should collect all datasets relevant to INSPIRE including services and metadata. Unfornutely, at least in my opinion, it's used more like trash can for all data sets which contain some part of data even distantly similar to those relevant to INSPIRE.

Searching of data and series is mediated through INSPIRE Discovery services. Discovery services are searching in metadata, specifically in `keyword` elements. Every provider can write into keywords anything he wants. That could be, and is, a problem. For accesing all data and services managed by COSMC, either by Section of central database or by Surveying Office, I conclusively recommend using the geoportal COSMC.

2. Implementation

During implementation of INSPIRE themes Addresses and Administrative Units, datasets and services harmonised by Implementation rules of INSPIRE Directive were designed, developed

---

[1]http://geoportal.cuzk.cz/
[2]http://geoportal.gov.cz/

and tested. Technical guidances for services and Data specifications were used during the implementation. Next step was making of metadata records. Metadata records serve as a description of data or services, not only human readible, but primarily computer readable.

I am personally engaged in a process of implementation since making of metadata records for data of the theme Cadastral Parcels. Themes Addresses and Administrative Units were implemented from the beggining to the very end with my participation. During implementation of themes Addresses and Administrative Units, metadata and data of the theme Cadastral Parcels were revised. This implementation took place in a few steps in the following order:

- analysis of Data specifications and Technical guidances INSPIRE,

- analysis of data in databases of COSMC,

- design of a data files structure,

- design of supported operations and planned limits for view and download services,

- creating of metadata records,

- testing and analysis of prepared data and services,

- revision of data files and services,

- revision of metadata,

- creating of promotional materials,

- publishing of data, services and metadata on the web `http://services.cuzk.cz/`,

- publishing of promotional materials on the web `http://geoportal.cuzk.cz/`.

In the future, Section of central database is going to continue in the implementation of INSPIRE Directive with the theme Buildings (BU). Concurrently with implementation of next INSPIRE theme, revisiones of already done themes are taking place in legislation. Revisions are based on experience and users feedback.

2.1. Data

Preparation of data is based on Data specifications on themes. Preparation of data is devided into three phases. In the first one, I have studied Data specifications on Addresses and Administrative Units. Second step was to analyse corresponding data in COSMC databases[3]. During the analysis it's necessary to decide which data from database are suitable to data structure according to the specification. For that purpose, I have made schemes of usage. In the third phase I have prepared sample file in GML 3.2.1 format. The sample file for each theme was sent to the firm Geovap, the developer of software `Marushka`®, which mediates generating of predefined GML files according to sample file for each theme.

The basic dispensing unit is different for each theme. For the theme Cadastral Parcels, there is one predefined file for each cadastral zoning. Addresses have one file for each municipality and all data for the theme Administrative Units are distributed in only one file for the whole

---

[3]ISKN – Information system of cadastre of real estates, ISÚI – Information system of territorrial identification, ZABAGED – Fundamental Base of Geographic Data

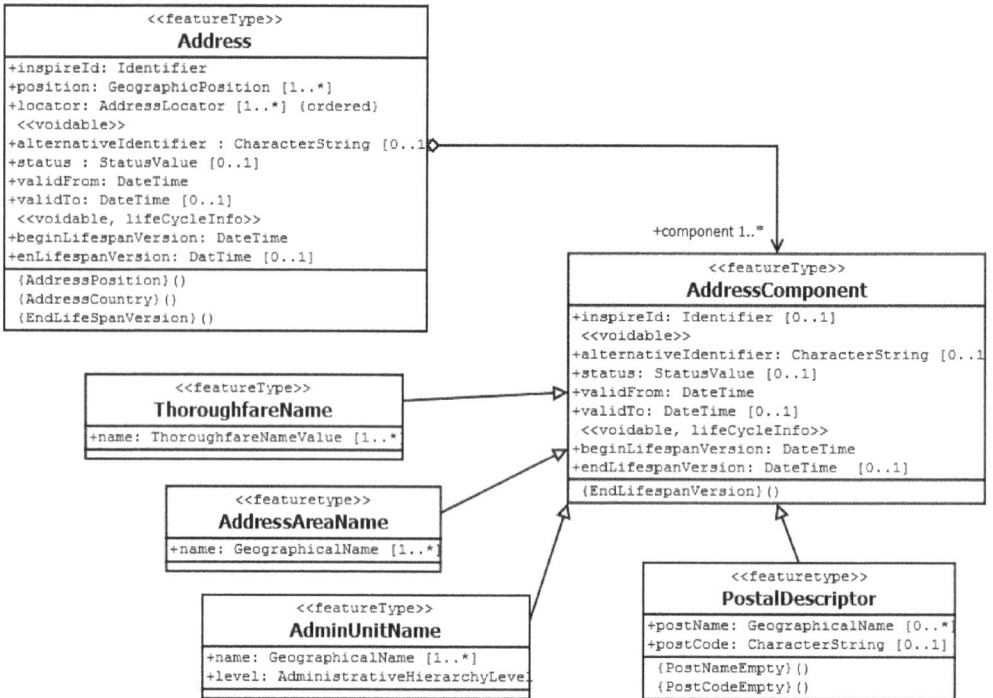

Figure 3: UML diagram of Application scheme on theme Addresses

Czech Republic. Predefined files are generated daily and are available for free on the page http://services.cuzk.cz/gml/inspire.

Marushka® software, besides providing of predefined files, also mediates INSPIRE harmonised download and view services according to the INSPIRE Technical guidance for services. These services are realised through OGC and ISO standards about WMS 1.3.0 and WFS 2.0.0.

2.2. Services

According to the INSPIRE Directive there is five types of services, which has to be provided for rightful implementation. These services shall be implemented:

- Discovery services – allow to search for data ad services according to keywords in metadata,

- View services – allow viewing data through Web Mapping Services in version 1.3.0,

- Download services – allow donwloading data through Web Feature Service in version 3.0 or through predefined GML files,

- Trensformation services – allow transformation of spatial data,

- Startup services – allow access for other types of services.

From the INSPIRE implementations point of view I was especially interested about implementation of download and view services, which allow direct access to the data. Data are continually updated in Publication Database. Sources of the data of Publication Database are ISÚI and ISKN. Data are essentially current, as the age of data two hours are featured.

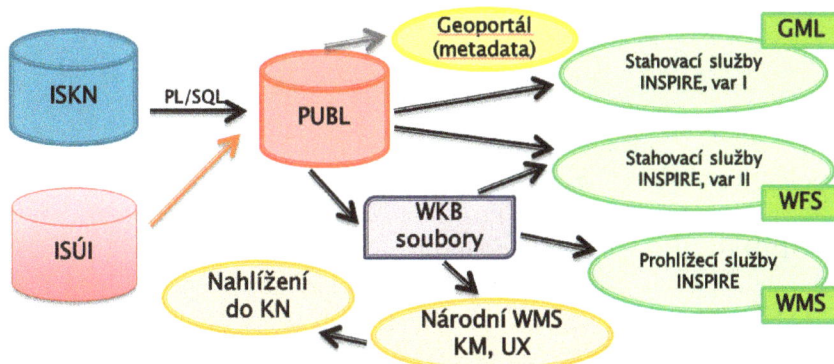

Figure 4: Process of managing data from databases ISÚI and ISKN for predefined GML files and WMS and WFS services *Source: Ing. Petr Souček, Ph.D.*

INSPIRE View services are realised through Web Mapping Service 1.3.0. Besides this version, older version 1.1.1 is also supported, but not forced by INSPIRE Directive. Access point for service is web address according to this model: `http://services.cuzk.cz/wms/inspire-[theme]-wms.asp?`. For example, view service for the theme Addresses has acces point on address `http://services.cuzk.cz/wms/inspire-ad-wms.asp?`. In order to simplify accessing data I have created a set of guidelines for using WMS services. Ther is one document for each theme:

- `http://services.cuzk.cz/doc/inspire-ad-view.pdf` – for Addresses,

- `http://services.cuzk.cz/doc/inspire-au-view.pdf` – for Administrative Units,

- `http://services.cuzk.cz/doc/inspire-cp-view.pdf` – for Parcels,

which contains list of available layers, supported coordinate systems and samples of requests.

Download services are realised according to Technical guidelines via WFS in version 2.0.0 and through predefined data files in GML format. Older versions of WFS aren't supported. Problem is, that WFS 2.0.0 is not supported by most software. Only one I know about, that supports this service is QGIS. Access is mediated through plugin WFS 2.0 written by Jürgen Weichand. Manual on downloading and using this plugin, including basic examples of working with it, is to found on this page: `http://services.cuzk.cz/doc/manual-wfs20-qgis.pdf`.

Access point for Web Feature Service is web page according to the model `http://services.cuzk.cz/wfs/inspire-[theme]-wfs.asp?`. Here's an example for Addresses: `http://services.cuzk.cz/wfs/inspire-ad-wfs.asp?`. Same as for WMS, for WFS I have created manuals too. They contain information about structure of data available through Web Feature Service and about the usage of this service. There is one document for each theme at the following addresses:

- `http://services.cuzk.cz/doc/inspire-ad-download.pdf` – for Addresses,

- `http://services.cuzk.cz/doc/inspire-au-download.pdf` – for Administrative Units,

- `http://services.cuzk.cz/doc/inspire-cp-download.pdf` – for Parcels.

Besides on-line access to data there is also a possibility to get a data through predefined GML files as described before.

2.3. Metadata

Metadata harmonised to INSPIRE has to follow Technical guideline for metadata. Its newest version (1.3) has been released on the 6th of November 2013. Metadata published by COSMC within INSPIRE is possible to divide into two parts. First one could be called "static", second one "dynamic". Static metadata include metadata for Series of INSPIRE datasets, metadata for INSPIRE Download services and metadata for INSPIRE View services. Dynamic metadata include getCapabilities documents for WMS and WFS, getFeatureInfo document, describeStoredQueries and other documents relative to network services. As the INSPIRE harmonised metadata are considered all metadata from the first category.

Technical guideline for metadata comes from technical norms ISO 19115 and ISO 19119 and National metadata profile and Metadata profile of COSMC also follow these norms. Metadata profile of COSMC includes everything what is required by INSPIRE Technical guidelines and National metadata profile and even more. Therefore I have used COSMC profile while I was creating metadata for INSPIRE themes.

All metadata has an identifier, which is unique in the scope of COSMC namespace. Combination of an identifier and namespace identifies metada record uniquely in the scope of the whole INSPIRE. Metadata describe service or metadata they are attached to. Besides description info they contain keywords. Keywords serves for discovering products through INSPIRE Discovery services. Every metadata record has a keyword according to GEMET thesaurus. For the data metadata, GEMET keyword serves as an identifier of the INSPIRE theme. Services metadata have an additional GEMET keyword which serves as an identifier of the type of INSPIRE service.

Other keywords should come from Vocabulary of COSMC, but not all INSPIRE related keywords are included. Currently we have initialized negotiations with Terminological commision about adding new keywords to the Vocabulary. Most of them are related to INSPIRE and Basic registers.

Metadata also include information about Data quality and its testing. For data and services, only tests used were on INSPIRE consistency and data completeness.

3. What's next?

By publishing data, metadata and services, implementation of INSPIRE isn't done yet. We have found a lot of mistakes and comments during implementation and I believe that so did most of European developers and analysts working on implementation of INSPIRE. Thats a reason why Maintanatce and Implementation Group (MIG) and Pool of Experts were founded. Ing. Jiří Poláček, CSc. is MIG member and both authors of this article are members of Pool of Experts. Implementation of INSPIRE is moving from the opening phase into the maintanance phase.

Within improvement of INSPIRE data and services it's really important users' feedback and continual development of data,, metadata and services. During my work on Addresses and Administrative Units I have revised metadata for Cadastral Parcels, which were published more than a year ago.

In the same time, interoperability of data and services between neighbour countries is going to be tested. Czech data and services are now tested together with Slovaks and cooperation with other neighbour states will follow.

References

[1] Poláček, J. Souček, P.: Implementing INSPIRE for the Czech Cadastre of Real Estates, Geoinformatics FCE CTU 8, 2012, pp. 9–16. [Online] [Cited: December 27, 2013.] `http://geoinformatics.fsv.cvut.cz/pdf/geoinformatics-fce-ctu-2012-08.pdf`

## 7

Geodetic surveying as a tool for discovering the prehistoric settlement in Sudan (the 6ᵗʰ Nile cataract)

Jan Pacina

J. E. Purkyně University in Ústí nad Labem
Czech Republic

Abstract

Surveying is an important part of any archaeological research. In this paper we focus on the archaeological research in north Sudan (6ᵗʰ Nile cataract) and the surveying methods applicable under the local conditions. Surveying in the Third World countries is affected by the political situation (limited import of surveying tools), local conditions (lack of fixed points, GNSS correction signal), inaccessible basemaps and fixed point network. This article describes the methods and results obtained during the three archaeological seasons (2011 – 2014). The classical surveying methods were combined with KAP (Kite Aerial Photography) to obtain the desired results in form of archaeological maps, detailed orthophoto images and other analyses results.

Keywords: Sudan, 6th Nile cataract, surveying, KAP, methods

1. Introduction

The Czech Institute of Egyptology has conducted research on archaeological concessions in Sudan (6ᵗʰ Nile cataract) since 2009 (Lisá et al., 2011; Suková – Cílek, 2012; Suková – Varadzin, 2012). This archaeological concession is located approximately 80 km downstream of Khartoum. It covers nearly 40 km of the west bank of the Nile and includes the whole western part of the Sabaloka Mountains and the zone is *ca.* 10 km in breadth extending from the riverbank towards the west and north-west (see Fig. 1). The interdisciplinary exploration of this area is aimed at a better understanding of the occupation of the Sabaloka Mountains and its vicinity during the Mesolithic and Neolithic periods (8ᵗʰ–4ᵗʰ millennia BC) and its interaction with the (changing) environment.

An integral part of the archaeological research is documentation by a variety of geodetic methods. The total station in combination with GPS measurements and Kite Aerial Photography was used for the surveying of the archaeological sites and features and topographic elements (such as terrain, settlements, paths). The results of the surveys are plans and archaeological maps, which supplement the satellite images of the research areas. Due to the absence of a fixed geodetic network and the GNSS correction signal, the survey was performed in a local coordinate system and even standard surveying procedures had to be adjusted from time to time to suit the local conditions. The objective of this article is to present the surveying methods applied in the extreme conditions of the Sudanese desert and the results achieved.

Figure 1: Area of interest overview (Sabaloka game reserve): 1 - the Fox Hill locality, 2 - the Sphinx Locality, 3 - prehistoric lake in the Lake Basin (basemaps: left – ESRI, right – Google).

2. Surveying conditions in Sudan

2.1. Importing surveying equipment

The Republic of South Sudan became independent in spring 2011. Since this period, the Republic of Sudan has applied very strict rules for the import of all surveying equipment (and many more things). All of the equipment (total station, GPS receivers, walkie-talkies) is subjected to special taxes. Permission issued by the Surveying Department in Khartoum allowing the usage of imported equipment may be required. The price of the permission depends on the type of imported equipment. Importing walkie-talkies (for long distance surveying) is not recommended as an additional permission from the Ministry of Communication is required.

2.2. Basemaps and satellite imagery

All of Sudan is, according to (Ali, 2009), covered by the Anglo-Egyptian Sudan Map Series, scale 1:250 000, created in the period of 1936-1951 by the United Kingdom Directorate of Overseas Survey. These maps are stored at the Survey department in Khartoum and they contain topographic layers together with approximate position of benchmarks and triangulation points. Other topographical maps available in our area of interest are maps 1:200 000, made by the Soviet Union in 1971. A more detailed map, is the one issued by the Sudan Survey Department in scale 1:100 000 – this map series was prepared in cooperation with the United Nations Development Program and it covers selected areas of the country (see Fig. 2).

There are no large scale maps available for the areas of interest, thus satellite imagery was used to identify terrain objects, possible settlement areas and for navigation purposes. The

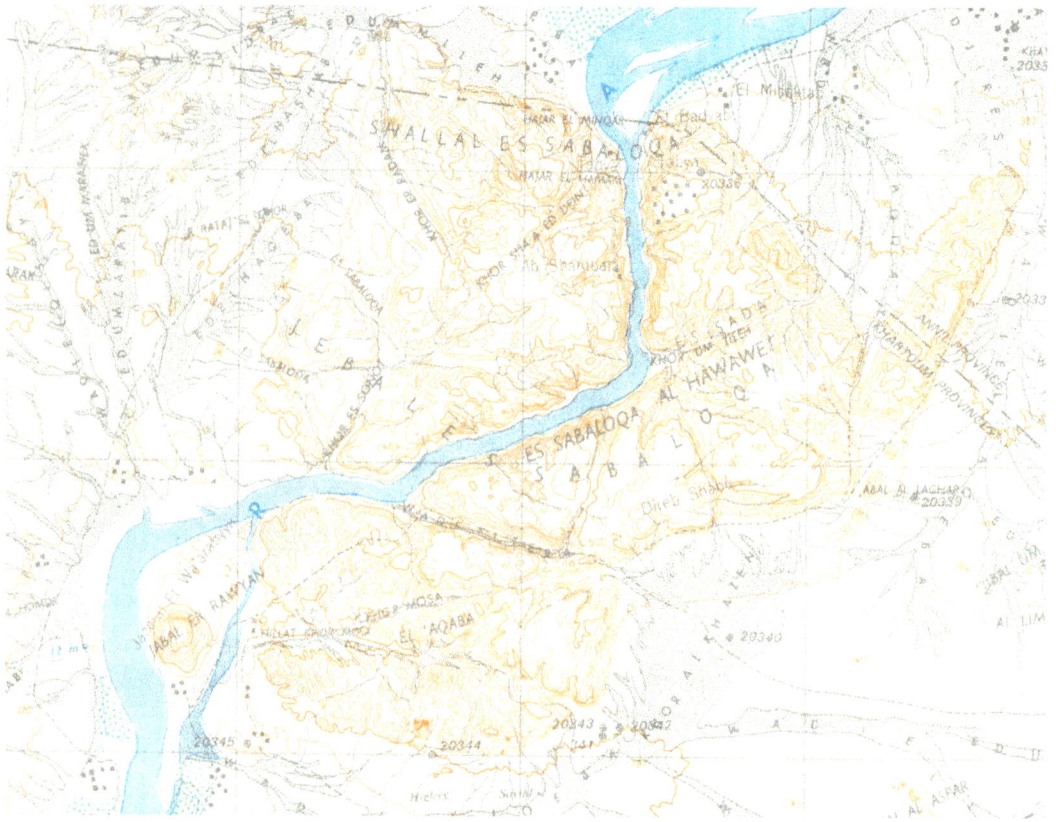

Figure 2: Area of interest presented on a 1:100 000 map.

WorldView 2 panchromatic (spatial resolution in nadir 0.46 m) and multispectral (resolution 1.8 m) imagery were chosen as a suitable data sources for research purposes. The data obtained, were further processed using the *pansharpening* method. With respect to the price of the data, only the core areas were covered by the WorldView 2 imagery at the Sabaloka site.

The entire area of the research site at Sabaloka (covers nearly 40 km of the west bank of the Nile and includes the entire western part of the Sabaloka Mountains and the zone ca. 10 km in breadth extending from the riverbank towards the west and north-west) was covered by satellite imagery accessible by Google Earth, handpicked and saved directly from the application. All the images were stitched together using a panorama stitching software into a seamless image. The produced image is not registered to any coordinate system.

Navigation in an unknown terrain required an easy way how to transfer registered satellite imagery into a rugged Garmin GPS receiver. The new Garmin BirdsEye Satellite Imagery was used in this case, it offered the same quality as it is accessible in Google Earth for our areas of interest. This was very helpful while moving within the Sabaloka mountains.

2.3. Coordinate system and geodetic control network

In Sudan, one can encounter many different coordinate systems – Adindan, 1942 or WGS84. Adindan is a geodetic datum suitable for use in Eritrea, Ethiopia and Sudan. Adindan references the Clarke 1880 (RGS) ellipsoid and the Greenwich prime meridian (GeoRepository, 2015). The 1942 coordinate system is used in the Russian maps 1:200 000 and WGS84 is used worldwide for the GPS NAVSTAR.

The coordinates of the archaeological sites were obtained from NCAM (Sudan National Corporation for Antiquities and Museums) in the Adindan coordinate system without any metadata. This coordinate system was identified in ArcGIS (ESRI corp.) as Adindan_UTM_Zone_36N and the data could be properly transformed into WGS 84 and used for navigation using a common Garmin GPS.

The geodetic control network at the Sabaloka locality was explored in the Survey department in Khartoum, but the closest benchmark and triangulation point on the west bank of the Nile is located in Obdurman (ca. 70 km upstream). Fixed points on the east of the Nile were closer, but located in a highly secured military area. Based on these facts, local coordinate systems were established.

2.4. The local geodetic control network

The absence of a geodetic control network at Sabaloka resulted in the creation of a local geodetic network covering all areas of interest – the total area where the surveying tasks were performed is larger than 400 ha. The terrain at Sabaloka is very complex, with hills, large rocky structures and flat desert planes making the surveying difficult.

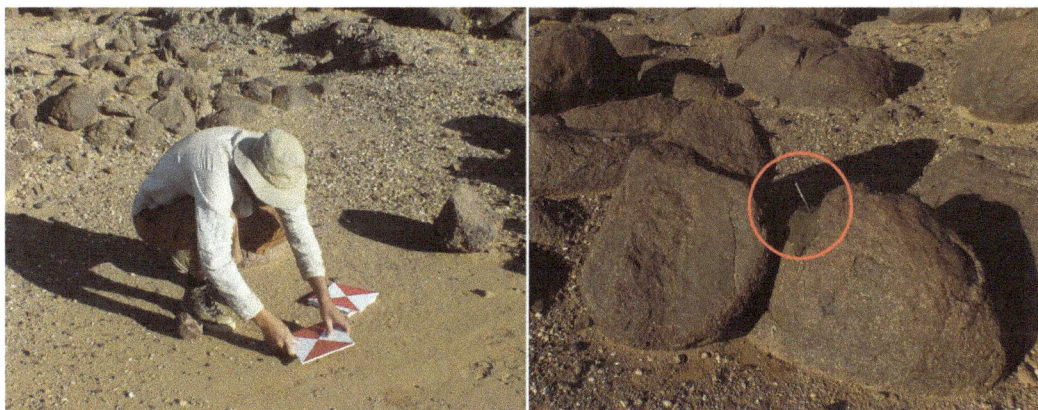

Figure 3: Points used as ground control points in the KAP survey (left), discrete fixed points (right).

The archaeological research begun in autumn 2011 at the Fox Hill locality (SBK.W-20/B) (Suková – Cílek, 2012; Suková – Varadzin, 2012) and here the beginning of the local coordinate system was situated. The coordinate system was initially meant to be used only for Fox Hill and the close neighborhood. The imprecise GPS measurements made it impossible to compare the elevation characteristics of the other archaeological sites within the Czech concession – thus the same coordinate system was "transferred" into all areas of interest.

Altogether 21 fixed points were established within the Fox Hill and the Sphinx (SBK.W-60) localities. All measurements performed in this area were based on these fixed points. The fixed point establishment was performed in 2012 by colour painting on selected rocks as the hills in the surroundings are made of friable granite. In the beginning of 2014, almost no fixed

Figure 4: Fox Hill locality and the surroundings.

points were identified as the extreme weather conditions destroyed the paint. The renewal of the local geodetic network was a crucial part of the 2014 expedition as incorporating the measurements from the past seasons with the current results was highly desirable.

3. Surveying instruments and tools

The most important imported surveying tool was the total station Leica TCR 303 (see Fig. 5). Two types of GPS receivers were used during the archaeological research – GPS Trimble Juno ST and Garmin GPSMap 62s. Despite of the lack of the GNSS correction signal, GPS usage was limited. GPS is mainly used for marking points of interest (settlement marks, important objects, terrain formations) discovered during the archaeological research in the wide area of the Sabaloka Mountains, marking the geodetic control network points as it is not easy to identify them in the rocky area, navigation to archaeological sites defined by NCAM and for navigation in the unknown terrain.

Figure 5: Surveying at the Sphinx locality (photo: L. Varadzin).

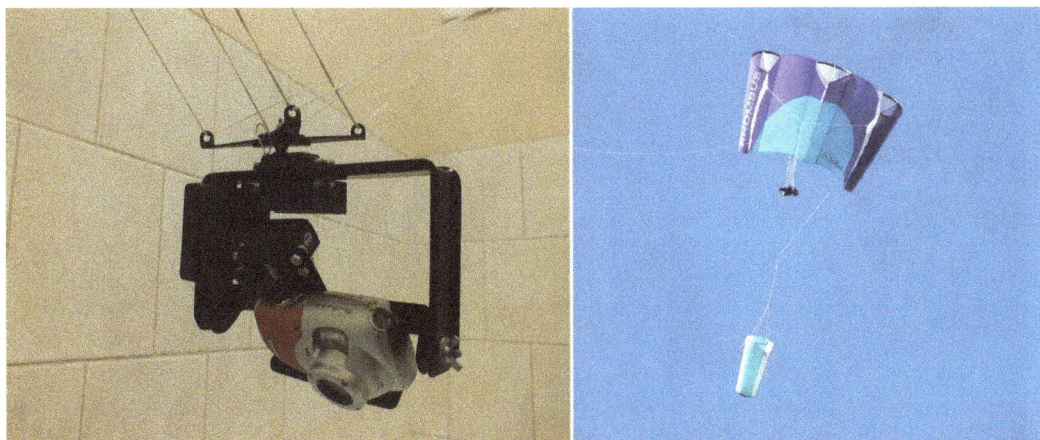

Figure 6: The Picavet suspension used on the Rhombus kite.

Kite Aerial Photography (KAP) was applied during the last archaeological season (autumn 2014). Small Format Aerial Photography is produced by KAP (Aber et al, 2010) which is usable for 3D models and orthophoto creation. KAP may be used as a supplement for UAVs (drones) in regions where their usage is forbidden. KAP has been successfully used in archaeological research in Africa many times (Bitteli, 2001; Brůna, 2013; Chagny – Hesse, 2007; Żurawski, 1993, 2005)

The one-string kite Elliot Rhombus Mega Power Sled, 300 × 170 cm, string length ca 200m, wind range 2 – 5 Bft and reinforcements GFK 2mm was used for the KAP survey. The same kite system was used with great success at archaeological sites at Abusir, Egypt (Brůna, 2013). The Picavet suspension (Verhoeven et al., 2009) is used to carry the camera used for SFAP (see Fig. 6). The camera holder was originally designed for a Canon Power Shot D10 camera with a mechanical trigger shooting images every 10 seconds. The 10 second shooting interval made this camera inappropriate for surveying large areas. A GoPro Hero 3+ camera was used instead. GoPro is a wide-angle, watertight camera with a 12MPix sensor and electronic shutter allowing time lapse images at a defined interval (2 second interval was used in this case). This camera is not primarily designed for SFAP (as it is equipped with a rolling shutter), but concerning its weight/pixel count/durability and based on our previous tests, this camera best served our needs in the KAP survey.

Other integral parts of the archaeological documentation are the simple measuring equipment which effectively support the surveying instruments. These are measuring tapes on a wheel of length 20 m and 50 m, tape rules (2 m or 5 m), folding rules (2 m), plummets, surveying stakes, strings for the delimitation of an archaeological squares, etc.

3.1. Surveying methods and precision

The total station Leica TCR 303 in combination with a surveying prism on a telescopic pole was used for the most of the mapping tasks. The range of the TCR 303 is maximally 2500 m and the longest sight lines within the archaeological mapping did not reach 1500 m. The total station is equipped with a program for computation of the free station method, measurements of minor geodetic points and indirect distance measurements. The methodology of "surveying

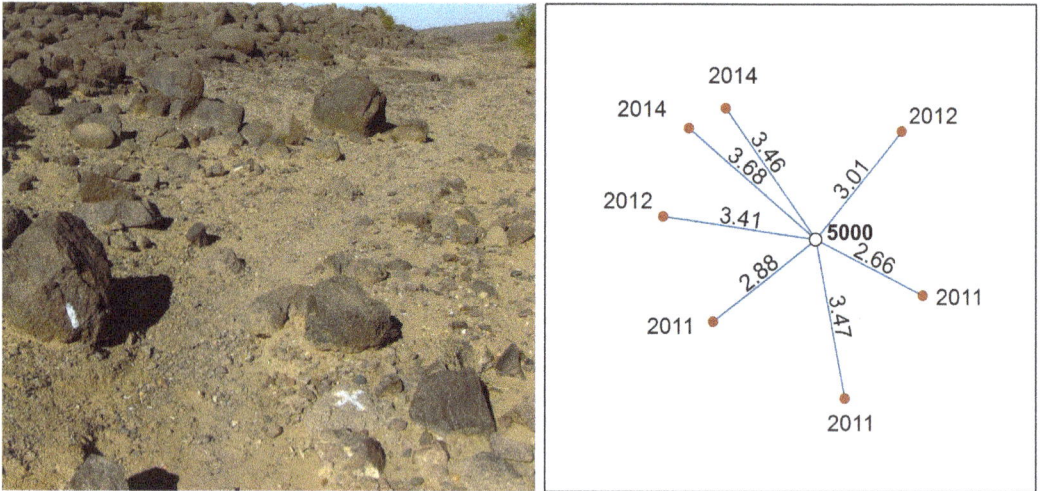

Figure 7: The initial point 5000 of the local coordinate control network (left). The control measurements on point 5000 during the archaeological seasons 2011, 2012 and 2014. The calculated distances are in cm (right).

in the desert" was many times proven during other archaeological expeditions in Egypt. See Brůna (2013) for more details.

The surveying conditions are much different from well-known European standards. The lack of a geodetic control network makes the surveying even more complicated and standard survey-ing workflows can't always be followed. Another factor is time stress – despite the expedition's time schedule and the surveying tasks time demandingness, all of the tasks had to be per-formed on the first attempt. This was a problem when making long-distance measurements. Long distance measurements are used while transferring the local coordinate control network to other archaeological sites in the area of interest. The measured distances were ca. 1000 m long and communication between the surveyor and the telescopic pole operator is problem-atic. Walkie-talkies normally used in Western countries, but are forbidden in Sudan – thus special flag signals (in combination with binoculars) are used to check whether the surveyor or the pole operator is ready to perform the measurement.

There are also several facts as well that may influence surveying precision. The imported total station may be stored in a custom warehouse for several days. Based on our experience, the goods stored in the custom warehouse were not always gently handled. The total station had to be checked and some basic calibration performed prior to surveying in the field.

Other facts influencing precision are the total station and the telescopic pole centering uncer-tainties - the fixed points are (with respect to the granite stones and rocks) marked by ca. 4 cm thick lines (see Fig. 7). Points established by the iron pipes (see Fig. 3, left) were built at the end of the 2014 season to preserve the points for further use. The vertical position of the telescopic pole is another crucial factor of the measurement but the set for the three-tripod system, suitable for long-distance measurements and fixed point establishment, is impossible to import into Sudan.

While transferring the coordinate system into other localities an oriented traverse is used.

Under standard conditions, such a long traverse should be oriented and connected on both ends to preserve the measurement precision. In the desert, we were forced to make the traverse only one side oriented and connected. Using this method will secure a sub-decimeter precision in the Z coordinate in-between the archaeological sites required for the Nile flood modelling. Transferring the coordinate system from Fox Hill to the Sphinx locality was the biggest surveying task. The traverse was almost 5 km long and 7 change points were used. This measurement lasted two days due to the rough terrain in which all the equipment couldn't be transported by a car.

The precision of the surveyed points was regularly checked by the control measurements on the established fixed points[1]. During every archaeological season, these measurements are performed to check the accuracy of the reconstructed fixed point network. An example is presented in Fig. 7, where the control measurements were performed on the initial point of the local geodetic control 5000 [1000, 1000, 377.87]. The measured coordinates are presented in Tab. 1. The RMS errors calculated based on these measurements are the following:

$$m_x = 0.020; \quad m_y = 0.021; \quad m_{xy} = 0.030.$$

Archaeologists required surveying precision (under these conditions) up to 5 cm, which was fully achieved.

Table 1: Point 5000 coordinates as measured during archaeological seasons 2011, 2012 and 2014.

Year	X 1000.000	Y 1000.000	Z 377.870
2011	999.977	999.982	377.876
2011	1000.023	999.988	377.863
2011	1000.006	999.965	377.876
2012	1000.019	1000.023	377.880
2012	999.966	1000.004	377.874
2014	999.980	1000.028	377.873
2014	999.972	1000.024	377.877

Another task was to compute the KAP derived models accuracy. The results derived from the KAP survey are Digital Surface Models (DSM) and orthophotos. Here we may want to know the accuracy of the resulting 3D model. The derived raster DSM is compared with the points surveyed during the 2011 and 2012 season (see the Fox Hill case study chapter). The resulting differences should evaluate the quality of the data. All of the tested DSMs have differences of no larger than +/- 5 cm from the reference data. The larger differences are caused by the sparse surveyed points which do not describe all of the rocks and stone formations in detail.

4. General conditions

The general conditions are much different from Western World standards. The whole archaeological expedition is living in the same conditions as the locals, accommodated in a rented

---

[1] While measuring archaeological data, the points of the (local) fixed geodetic network in the surroundings are measured as unknown points to evaluate the measurement accuracy.

house near the archaeological site. The house is built of bricks dried in the sun and usually contains of two large rooms, a hall and a large back-yard (used as an open-space bedroom). Food is obtained from local sources and prepared by a hired Sudanese cook. Nile water is used for washing and cooking and bottled water for drinking. Electricity has been available since 2013, so the diesel generator used for charging all the equipment is no longer used.

Sunrise is about 6am and sunset about 6pm. Work starts usually at 7am and there is a lunch break from 12.30 pm to 3pm, as the weather is very hot during midday. High temperatures (over 50°C) are not suitable for the surveying equipment. Stronger winds sometimes bring dust and sand in the air and these conditions are not suitable for long distance measurements.

The local people are mostly friendly and hospitable, but the fixed points have to be marked in a very discrete way as a cross made on the stones by the *foreigner* means that there is gold hidden underneath. This has led to the destruction of fixed points on several occasions.

5. The Fox Hill locality (SBK.W-20/B)

The Fox Hill locality is the most important prehistoric site located within the Czech archaeological concession in the Sabaloka region. The site is structured on 16 terraces and platforms, the total surface area of which is 11,650 m^2 (Fig. 8). The terraces and platforms are well delimited by the exposed bedrock and boulders and vary in size, elevation and ease of access. The settlement is dated back to the Mesolithic (ca 9000 – 5000 BC) and Neolithic (5000 – 3000 BC) periods (Suková – Varadzin, 2012).

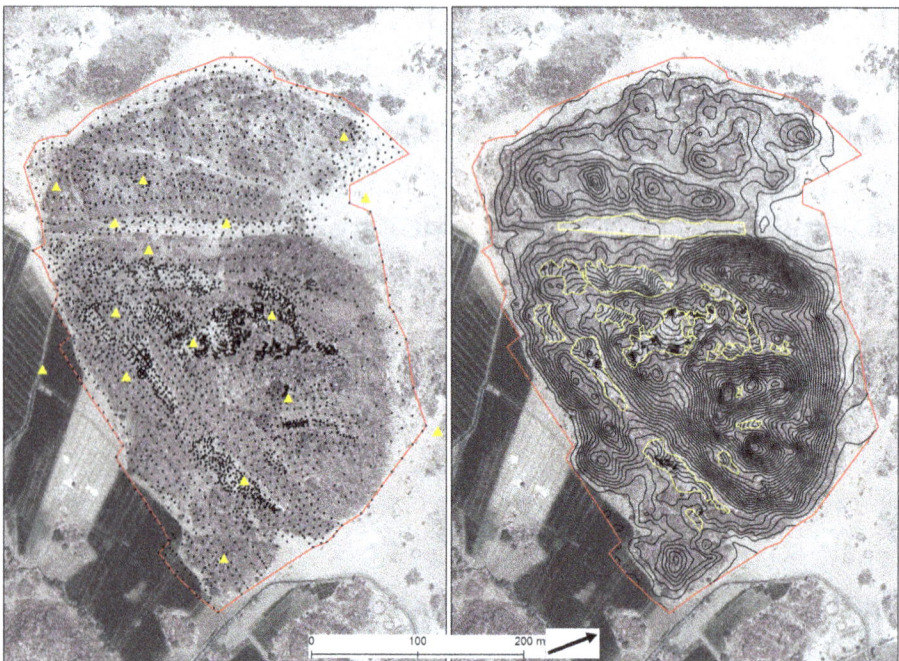

Figure 8: Surveyed points at the Fox Hill locality with marked fixed points (left). Contour lines generated from the DEM generated based on the surveyed points and the delineated occupation terraces (right). (Basemap: Google)

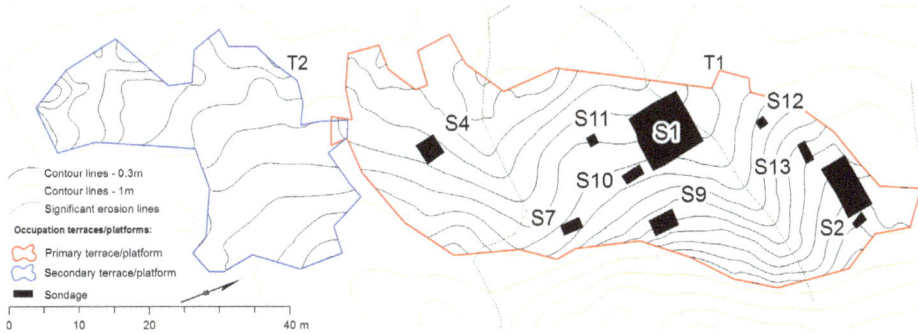

Figure 9: Detail of Terrace 1 and Terrace 2 at the Fox Hill locality.

At Fox Hill, 17 geodetic control points were established and more than 4500 points were measured. The occupation terraces are covered by a 1x1m point network and the rest of the area by a 5x5 m network, supplemented by points of erosion lines, ridges and significant stone structures.

Every site has its own specific environment, thus the choice of a proper interpolation method is required (Karel, 2006; Pacina, 2013). Based on previous results and tests, interpolation methods implemented in the ArcGIS (Topo to Raster) were used for producing a DTM of the localities.

Contour lines of a 0.3m interval for the occupation terraces and a 1m interval for other areas were derived from the interpolated DTM. See Fig. 8. The contour map may describe the spatial relationships between adjacent areas in the locality, material transportation caused by erosion (the archaeological soundings started in the areas that are not affected by erosion), and possible connections between the settlement areas and their elevation.

Delimitation of the occupation terraces at Fox Hill must be done by an archaeologist, as they recognize the boundaries of the terrace. An example of the delineated terrace, together with archaeological soundings is shown in Fig. 9. Fig. 11 shows objects excavated at Terrace 1 (T1), Sounding 2 (S2) in detail.

6. The Sphinx locality (SBK.W-60)

The Sphinx locality, positioned on a granite outcrop in the Rocky Cities area is the second most important site in the Czech concession. The outcrop features only one platform of 940 m^2 situated ca 15 m above the surrounding terrain. The surface finds at Sphinx attest to the occupation of this site only during the Mesolithic period (Suková – Varadzin, 2012).

Despite of the distance between the two localities it was decided to make a local coordinate system at Sphinx as well. The archaeologist requested a sub-meter precise comparison of the altitudes of these two localities in connection with the possible local flooding by Nile. A test using the Garmin GPS receiver was performed to define the altitude of the two known points previously measured with the total station. Altitude calibration was performed on one of the points and the other point several hundred meters away was then measured. The result was unsatisfactory – the difference in the altitude was almost 10 m. It was decided to put the Sphinx locality in the same local coordinate system as Fox Hill. This was performed by a *one*

Figure 10: Fox Hill - Terrace 1, Sounding 2 - excavated circular objects.

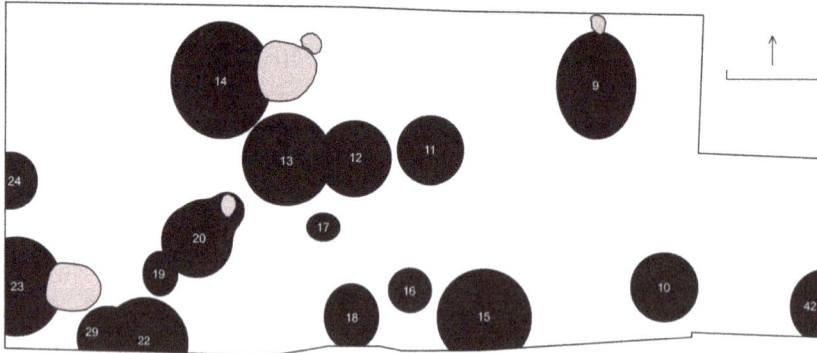

Figure 11: Objects excavated at Terrace 1, Sounding 2 – compare with Fig. 10.

point oriented traverse (7 points, 4.7 km length). Based on previous precision tests performed in this area; we may consider the precision of the applied method satisfactory.

Within the Sphinx locality, about 400 points on settlement terrace of ca. 900 m^2 (see Fig. 12) were surveyed. The resulting archaeological map is presented in Fig. 14 and the sounding excavated in season 2012, in Fig. 13. The archaeological map has a detailed orthophoto used as the basemap. This orthophoto was created using KAP and the Small Format Aerial Photography. Altogether more than 1300 images were taken during the KAP survey. The camera on the kite swung and thus only a small amount of the imagery could be considered horizontal and used for further processing. Only 60 images were used for creating the 3D model of the Sphinx locality and the derived orthophoto (presented in Fig. 14). Image processing was per-

Figure 12: The Sphinx locality with marked occupation terrace.

Figure 13: The Sphinx locality - research at Sounding 2 (2012).

formed using PhotoScan (Agisoft LLC) software. Detailed processing workflow is described i.e. in (Verhoeven et al., 2012).

7. Discovering a prehistoric lake – the Lake Basin locality

The Lake Basin area adjoins the massif of Jebel Sabaloka from the south west and extends over ca. 2.2 km x 1.5 km. This microenvironment is closely connected to the Nile, the waters of which used to submerge the lower reaches of this area during the annual floods in prehistoric times. This landscape type is formed of more than 20 mostly granite rock outcrops which vary in their size and in the number of natural terraces and platforms suitable for occupation (see Fig. 15). The former occupation sites are arranged around a depression where the remains

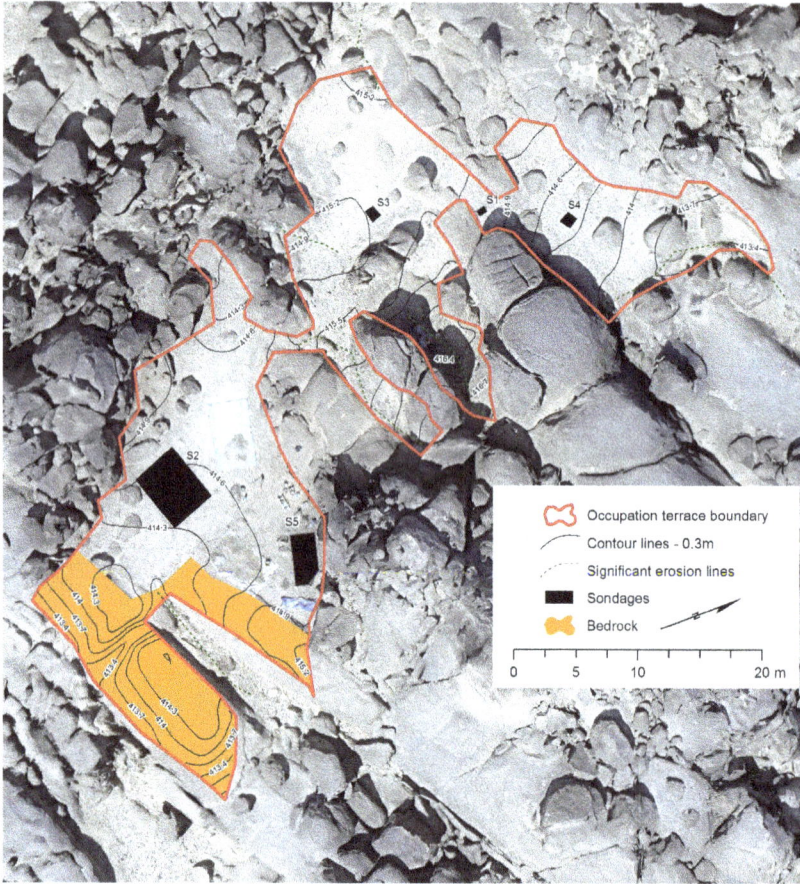

Figure 14: Archaeological map of the Sphinx locality.

Figure 15: Lake Basin with granite rock outcrops.

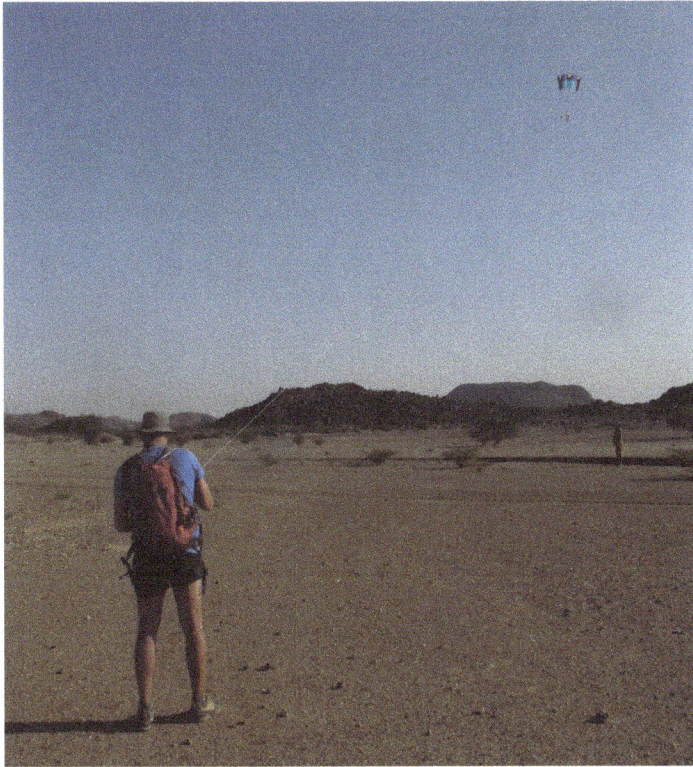

Figure 16: KAP in the Lake basin (Photo: J. Novak).

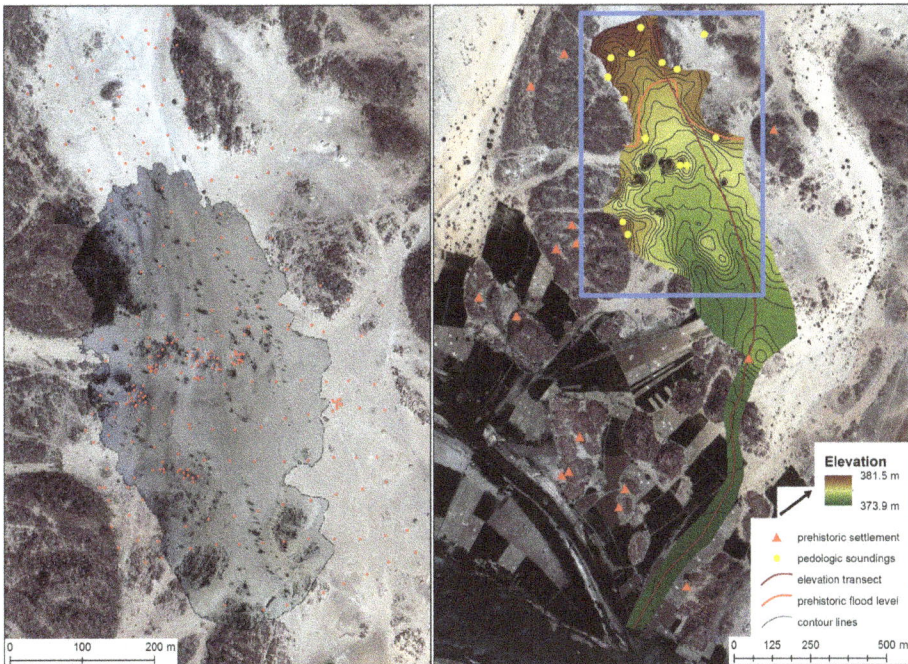

Figure 17: Point surveyed within the Lake Basin area and orthophoto derived from the KAP survey (left). The created DTM of the prehistoric lake bottom with locations of identified archaeological sites and sediment soundings. (Basemap: Google)

Figure 18: Orthophoto image created based on the KAP survey – marked sounding at the tumulus foot.

of a prehistoric lake were identified (Suková – Varadzin, 2012).

Since the 2012 season, it was desired to create a 3D model of the prehistoric lake bottom and prove the relationship with the Nile water level and potential prehistoric flooding. With respect to the natural conditions in the Sabaloka region, it is not easy to visually estimate elevations and distances. The lake bottom reconstruction has been done using several methods. In 2012, two transects were surveyed proving that this area could be flooded by the Nile. More effort was used to work on this topic in the 2014 season. The whole bottom of the potential lake (ca. 25 ha) was covered with a regular 25 x 25 m point grid and surveyed with the total station densified in the areas with a more complex terrain.

The KAP survey was used in the areas where the sediment samples were taken. The area of interest was surveyed twice as during the first attempt a very strong wind caused an uncontrollable kite fall resulting in camera damage. The second attempt went well, but it was proven that the KAP method is not suitable for covering such large areas, using the equipment used so far. The camera swings on the kite and a longer observation at one point is required to obtain horizontal images – this can be achieved in smaller localities but impossible at a 25 ha area. The oblique images in different directions "break" the point cloud computed by the Structure from Motion algorithms. More than 1600 images were taken during this survey and only 200 were finally used for the 3D model and orthophoto creation.

The area surveyed by the KAP method and covered with surveyed points is presented in Fig. 17. Based on all the obtained data, the DTM representing the former lake bottom was created. In Fig. 17 (right) the prehistoric settlement sites (occupation terraces) in the Lake Basin surroundings are identified. The altitude of these sites varies from 378 to 380 m.

Based on field research, it was determined that these settlement sites were (probably) not flooded during the seasonal Nile floods and thus this elevation value was used for modelling the level of prehistoric floods. Details of the resulting orthophoto image of the Lake Basin (with 3 cm/pixel resolution) are presented in Fig. 18. In the figure, the soil profile containing the original prehistoric lake sediments, protected from erosion by the tumulus is marked.

8. Conclusions

The modern methods of spatial data collection offer fast and quality data. Surveying in developing countries is a little different as new methods can't be imported (UAV) or properly used (GNSS correction signal). The surveying methods applied in the Sudanese desert are not absolutely fitting into Western standards but still offer the best possible results under these extreme conditions. In this paper, spatial data collection methods are presented used within the archaeological research that offer a new point of view on archaeological sites and findings. One interesting data collection method is Kite Aerial Photography that is used as an UAV supplement. We were capable of creating orthophoto and 3D models of the sites. 3D models created during the 2014 season are presented at https://sketchfab.com/jan.pacina/folders.

References

[1]   James S. Aber, Irene Marzolff, and Johannes B. Ries. *Small-Format Aerial Photography: Principles, Techniques and Geoscience Applications.* Amsterdam – London: Elsevier Science, 2010. ISBN: 978-0-444-53260-2.

[2]   Abdullah Elsadig Ali. "Current Status of GIS in the Sudan". In: *Eighteenth United Nations Regional Cartographic Conference for Asia and the Pacific, Bangkok.* United Nations E/CONF.100/CRP.10, Oct. 2009.

[3]   G. Bitelli, M. Unguendoli, and L. Vittuari. "Photographic and photogrammetric archaeological surveying by a kite system". In: *Proceedings of the 3rd International Congress on Science and Technology for the Safeguard of Cultural Heritage in the Mediterranean Basin.* Ed. by Jesús Alpuente et al. Alcala de Henares: Servicio de Publicaciones de la Universidad de Alcala, July 2001, pp. 538–543.

[4]   Vladimír Brůna. "Využití KAP (Kite Aerial Photography) při dokumentaci výzkumu v Abúsíru". In: *Pražské egyptologické studie, UK Praha* XI (2013), pp. 37–43.

[5]   B-N. Chagny and A. Hesse. "Soudan 1994–2006: Photographies archéologiques sous cerf-volant avec Francis Geus". In: *Mélanges offerts à Francis Geus. Égypte-Soudan, Lille: Université Charles de Gaulle-Lille.* Ed. by Brigite Gratien. Cahiers de recherches de l'Institut de papyrologie et d'egyptologie de Lille 26, pp. 47–59.

[6]   *GeoRepository, Geodetic Datum used in Africa - Eritrea, Ethiopia and Sudan.* http://georepository.com/datum_6201/Adindan.html. Oct. 2015.

[7]   W. Karel, N. Pfeifer, and C. Brueser. "DTM quality assessment". In: *ISPRS Technical Commission II Symposium, Vienna.* Ed. by W. Kainz and A. Pucher. 2006, pp. 7–12.

[8]   Lenka Lisá et al. *Sabaloka a Šestý nilský katarakt.* Novela Bohemica, 2012. ISBN: 978-80-904573-6-2.

[9]   J. Pacina et al. "Detailed analysis of georelief development in the lake Most surroundings". In: *Ad Alta: Journal of Interdisciplinary Research* 3.2 (2013), p. 44.

[10] L. Suková and V. Cílek. "The Landscape and Archaeology of Jebel Sabaloka and the Sixth Nile Cataract, Sudan". In: *Interdisciplinaria Archaeologica – Nat. Sciences in Arch.* 3.2 (2012), pp. 189–201.

[11] Ladislav Varadzín and Lenka Suková. "Preliminary report on the exploration of Jebel Sabaloka (West Bank), 2009-2012". In: *Sudan & Nubia* 16.1 (2012), pp. 118–131. ISSN: 1369-5770.

[12] Geert J. J. Verhoeven et al. "Helikite aerial photography - a versatile means of unmanned, radio controlled, low-altitude aerial archaeology". In: *Archaeological Prospection* 16.2 (Apr. 2009), pp. 125–138. DOI: 10.1002/arp.353.

[13] G. Verhoeven et al. "Mapping by matching: a computer vision-based approach to fast and accurate georeferencing of archaeological aerial photographs". In: *Journal of Archaeological Science* 39.7 (July 2012), pp. 2060–2070. DOI: 10.1016/j.jas.2012.02.022.

[14] Bogdan Żurawski. "Low altitude aerial photography in archaeological fieldwork: the case of Nubia". In: *Archaeologia Polona* 31 (1993), pp. 243–256.

[15] Bogdan Żurawski. "Miracles of Banganarti. Polish archaeological discoveries in Sudan". In: *Focus on Archaeology* 5.1 (2005), pp. 20–23.

## 8

Database for tropospheric product evaluations - implementation aspects

Jan Douša, Gabriel Győri

Research Institute of Geodesy, Topography and Cartography,
Geodetic Observatory Pecný
Ústecká 98, Zdiby 250 66
jan.dousa@pecny.cz

Abstract

The high-performance GOP Tropo database for evaluating tropospheric products has been developed at the Geodetic Observatory Pecný. The paper describes initial database structure and aimed functionality. Special focus was given to the optimizing effort in order to handle billions of records. Evaluation examples demonstrate its current functionality, but future extensions and developments are outlined too.

Keywords: troposphere, zenith delays, database, GNSS, radiosonde, meteorological data

1. Introduction

The potential of GNSS observations for troposphere monitoring has been described in Bevis et al. (1992). Since that time various projects aimed for developing GNSS-meteorology in Europe. First benchmark of near real-time ground-based GNSS tropospheric products – Zenith Total Delays (ZTD) – was provided within the COST Action 716 – Exploitation of Ground-based GNSS for Meteorology and Climatology (1999-2004). The extended routine analyses were supported by the EU FP5 project – Targeting Optimal Use of GNSS for Humidity (TOUGH, 2003-2006). Recently, the E-GVAP I-III (2006-2016) aimed for the establishing operational ZTD estimations and their assimilations in numerical weather models (NWM) and for developing active quality control for GNSS products. Additional effort on enhanced capability of GNSS troposphere monitoring and the exploitation of NWM data for GNSS positioning has been recently prepared and approved within the COST Action ES1206 – GNSS for Severe Weather Events and Climate (GNSS4SWEC, 2013-2017).

The Geodetic Observatory Pecný (GOP) analysis centre has contributed to the above projects since 2000 and provided one of the first operational GPS tropospheric products - near real-time regional GPS solution available officially since 2001 (Douša, 2001). Additional tropospheric products in support of meteorology have been developed at GOP during recent years – near real-time regional multi-GNSS product, 2011-present (Douša, 2012a), first near real-time global product, 2010-present (Douša 2012b) and real-time ZTD product, 2012-present (Douša et al., 2013). GOP routine post-processing tropospheric solution has been available also since 1996 for the part of the EUREF Permanent Network (EPN). The complete European network was reprocessed at GOP recently for the entire period 1996-2012. As being the most accurate and homogeneous during the whole interval, the reprocessing could be used in regional studies for climatology. The regular and long-term evaluation of all these products is an important task for both getting a relevant feedback about the accuracy and studying potential for improvements.

Initial comparison of GOP tropospheric products was done occasionally (Douša, 2003) using

PERL scripts for data stored in internal text format. With increased data period such design was recognized as inconvenient and it was replaced by a simple MySQL database used during 2002-2010 for GOP GNSS-based zenith total delays routine evaluations. Recently, more flexible database structure was requested for fully automating tropospheric data comparisons including the searching of nearby points, filtering, converting, interpolating, generating various statistics and their extracting for web-based plots. New database (labelled as 'GOP Tropo database') was required to provide a high-performance system in order to deal with billions of data records. As a free alternative to the enterprise solutions, the PostgreSQL server (POSGRESQL) was selected for this task and the database structure was completely revised in order to support its flexible utilization. The GOP Tropo database structure and functionality design is described in Section 2, the optimization aspects in Section 3 and initial comparison examples in Section 4. The conclusion and outlook is given in the last section.

2. Database design

This section provides a rough description of the data structure with the focus on specific details. The data organization is the most important aspect of any database since it predefines any future utilization (and its flexibility for extensions). The main GOP Tropo database features and functionalities were initially defined as well as requirements for future developments to:

- accommodate and compare different tropospheric data types in a single database,

- provide geo-referencing and collocated point searching for data comparisons and interpolations,

- apply vertical corrections (potentially supported with geoid and orography models),

- generate comparison differences and statistics in a yearly, monthly, weekly and hourly mode,

- support conversion data types (e.g. zenith wet delay to integrated water vapour and others in future),

- interpolate values from grid points and calculate grid points from available data (in future),

- study trends, temporal/spatial variations, correlations (in future).

Target and potential utilizations are foreseen such as a) to compare (near) real-time GNSS tropospheric products with respect to the final ZTD products, b) to assess GNSS results with respect to radio soundings, radiometers or numerical weather models and other independent datasets, c) to compare troposphere estimates from different space geodetic techniques (GNSS, VLBI, DORIS), d) to evaluate different strategies of interpolations or kriging of meteorological data or tropospheric delays, e) to evaluate in-situ meteorological observations provided in M-RINEX, f) to evaluate global pressure, temperature or specific tropospheric models in future.

2.1. Database structure

In any relational database the user data are placed in database tables, which structure is designed optimally for the purpose of the utilization. The data are organized as records (rep-

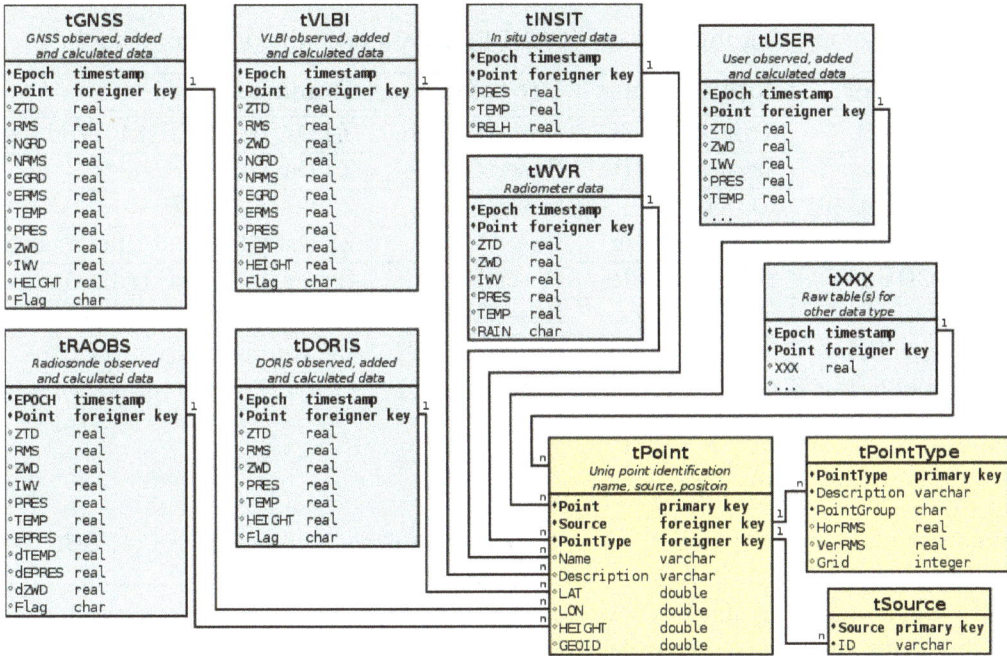

Figure 1: Basic structure of GOP Tropo database including original data representation.

resented by table rows) while each record has specific values (represented by table columns). The data should not be duplicated and, when any duplication occurs, new database table with relevant records is created to unify any dualities. Such unique record is then related to the original table record by using the relation provided with a reference key (that's why such database is called relational).

The GOP Tropo database is designed to accommodate different tropospheric meteo data types such as: GNSS, VLBI, DORIS, radiosondes, radiometers, synoptic data, in-situ meteorological data, data extracted from the NWM and other supporting data/models as e.g. geoid, orography. According to the variability of tropospheric products, meteorological data or other supporting data, it would not be easy and efficient to accommodate them within a single data table, because each data source has its own specific stored values. Fortunately, data from each source could be processed independently and it was identified as useful to define a specific data table for each data type. Different data sources for all data types are also considered, e.g. such as the GNSS ZTDs from EUREF, E-GVAP analysis centres, the International GNSS service and many others. Additionally, for a single analysis centre, e.g. GOP, various products are available too, such as final, reprocessed, near real-time (global, GPS, GPS+GLONASS), real-time or others. All the sources within a single data type are accommodated within a single table providing unique source identification for the data.

The common structure of the data organization within the database is shown in Fig. 1. All data are georeferenced according to the reference key to the *tPoint* table ('t' is always used to identify the table name), which provides additional information about the data source (*tSource*), site identification (*tSite*) and point location. Optionally geoid undulation and orography height is provided too. Each record in the *tPoint* table is uniquely defined by its name, source and position (latitude, longitude, height) along with the position accuracy used

Table 1: List of existing input filters (PERL) for data decoding and inserting in database

Input filter	Input format	Procedure	Remarks
tro-snx2DB.pl	Tropo-SINEX	fInsertGNSS	ZTDs, IGS/EUREF products
bsw-trp2DB.pl	Bernese TRP	fInsertGNSS	ZTDs, GOP products
cost-trp2DB.pl	COST-716	fInsertGNSS	ZTDs, E-GVAP
rt-flt2DB.pl	Tefnut output	fInsertGNSS	ZTDs, GOP real-time analysis
raobs2DB.pl	BADC profiles	fInsertRAOBS	Integrated data, radiosondes
wvr2DB.pl	Radiometrics	fInsertWVR	Integrated data, radiometers
met-rnx2DB.pl	Meteo RINEX	fInsertINSIT	In-situ meteo data (GNSS)
cost-met2DB.pl	COST Meteo	fInsertSYNOP	COST 716 meteorological data

in identification of a unique point.

Specific data tables are currently predefined for the data types – *tGNSS, tVLBI, tDORIS, tRAOBS, tWVR, tINSIT, tUSER, tGEOID, tSURF* and others where the name can help to identify the data content. Others specific data could be completed later, such as for mapping functions coefficients, slant tropospheric delays, synoptic data, NWP grid data (or more likely their reduction to the specific parameters at a reference surface) and other.

2.2. Database feeding, record uniqueness

The database filling is done in three steps: 1) data download, 2) decoding and converting original data including data preparation for SQL command calling a specific database insert procedure and 3) executing SQL command within GOP Tropo database. Input data are collected from various sources via the standard ftp or http downloads to a local disk, all in original and usually compressed formats. This process is controlled via unix cron job scheduler.

The data decoding and filtering is done by in-house developed input filters written in the PERL scripting langue. Their main tasks consist of the extracting and converting of data from text files and preparing (and executing) SQL commands calling the GOP Tropo database insert procedure. The input filters are designed for various input formats (e.g. for *tGNSS* it is Tropo-SINEX format, Bernese 'TRP' output, COST-716 format) and also specific database insert procedure is called for each data types (e.g. *tGNSS* uses *fInsertGNSS* procedure). The list of supporting formats and input filters is given in Tab. 1. The radio sounding profile is the example of data type, content of which is not fully included into the database, but only selected parameters derived at the surface (e.g. pressure, temperature, water vapour pressure, zenith hydrostatic delay, zenith wet delay, relevant lapse rates for possible vertical conversions etc.). The data decoding and database filling is also regularly started from a cron scheduler.

The internal database stored procedure for inserting is performed either via INSERT or UPDATE SQL command. The former command supports direct inclusion, but works for new records only. The latter consists of an initial check if the record already exists in the database and, if true, it replaces it with the current data. In this context it is important to understand how records are identified as unique within data tables. Database systems use primary keys to specify the column (or columns) that uniquely identify each record, and these can be either natural or surrogate. A natural key is the one that is composed of columns directly related to the data, while surrogate key is a specific column added to the data table only for serving as a primary key (e.g. a unique identifier, auto-incremented numerical value). In many cases we use surrogate keys, but not for the main data tables, where the primary key is designed as a unique index over two columns – *Epoch* and *tPoint*. The epoch value is of the type 'TIMESTAMP WITHOUT TIMEZONE' handling commonly data and time. The point column refers to the record in the *tPoint* table, where the uniqueness is provided via auto-incremented numerical value. The record in the *tPoint* table is checked and updated automatically anytime when filling new entry into main data table. This is proceeded as follows – requested point is searched within the *tPoint* table and, if found, it returns the reference key and, if not, the new point is created. The uniqueness of the point records is thus managed within the point searching/inserting internal procedure (*fInsertPoint*) taking into account the unique point characteristics: *SiteID, Source, PointType* and *location* (latitude, longitude, height) within predefined *point accuracy* (horizontal and vertical).

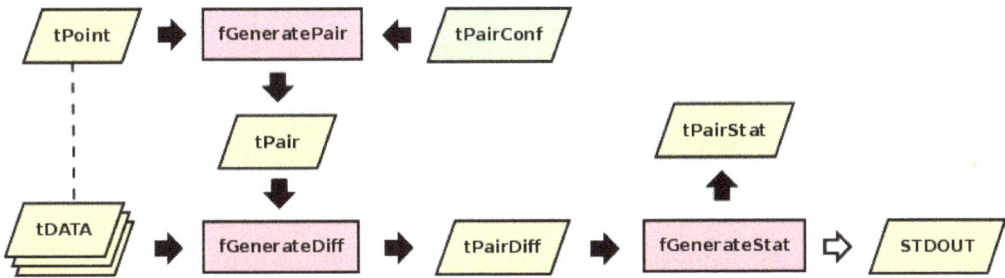

Figure 2: Scheme of the comparison within GOP Tropo database

2.3. Data and product comparison

The main functionality of the current GOP Tropo database implementation is the comparison of various tropospheric data and products. It consists of this sequence of steps (Fig. 2):

- comparison configuration (manual),
- search of collocated points (with respect to the setting criteria),
- generation of data differences for identified pairs,
- statistics over data differences,
- extraction and visualization (provided outside the database).

The user configures data or product comparisons by defining two data sources from one or two data tables. The setting consists of criteria for searching pairs of collocated points – the limit for the horizontal and vertical distance of two points. If applicable, the maximum σ is set for the initial data filtering and the limit of the confidence interval is provided for the statistical procedures to detect and reject outliers. Optionally, mask or explicit site list can be provided if selected stations should be compared only (implicitly all stations). Individual comparison settings are stored as records in the special database table (*tPairConf*) where a surrogate key is set for further referencing.

The candidates for identifying collocated points are searched within the *tPoint* table for the two specified sources from the setting. The station pairs are generated by the *tGeneratePair* procedure and the pair list is stored in the *tPair* table used afterwards for data differences generation (*fGeneratePairDiff*) within the period of request. Compared products have usually different sampling rates and the values for comparing could be generally referred to different epochs. For this case, the database supports sampling rate argument defining the interval within which all single product values are extracted and the mean value is calculated for the comparison. An optimum difference sampling step should be set up as the higher data sampling of both products in order to grab common values and, on the other hand, below one hour since the variation of the troposphere will be significantly smoothed by averaging product values over a longer time. The setup between 10 and 60 minutes is usually reasonable. In future, we consider supporting other functional fitting of the single product values for differencing instead of a simple averaging. Generated data differences are always stored in the *tPairDiff* table enabling to accommodate differences from recently collected data or analyzed products.

The statistical procedure (*fGeneratePairStat*) applies three iterations to estimate biases, standard deviations and root mean squares (r.m.s.) for each compared pair individually. The first iteration serves for the median estimation as a robust initial mean value. It is used in the second step for calculating differences reduced by mean value in order to estimate reliable standard deviation for the outlier detection. In the third step, final statistics – bias, standard deviation, r.m.s., number of all observation and outliers – are calculated after outliers were rejected using the confidence interval based on standard deviations from the second iteration and, optionally, data excluding r.m.s. limit from the setting. For individual comparisons, five statistic modes are provided for a period of request (defined by 'beg' and 'end' arguments) – all, yearly, monthly, weekly and hourly. The first one calculates the statistics over the requested period, the last for the same period but providing statistics for data filtered according to hour of day. All other modes calculate statistics individually for windows as specified within the whole period. Because the differences are saved in the *tPairDiff* table, all statistic modes can be efficiently repeated in a regular update for generating time-series as demonstrated in Fig. 3. Finally, the extractions and visualizations of results are described in the section with sample evaluations. Table 2 summarizes the comparison procedures. Note that each comparison procedure uses a key to the configuration record in the *tPairConf* as its first argument which has been omitted in the table.

Table 2: List of comparison procedures/functions ('f') input/output tables ('t')

Procedure	Arguments	Input	Output	Remarks
fGeneratePair		tPairConf	tPair	generate pairs
fGeneratePairDiff	'beg', 'end', 'sample'	tPair,tData	tPairDiff	generate differences
fGeneratePairStat	'beg', 'end', 'type'	tPairStat		generate statistics

2.4. Other procedures for data maintenance

Since data are structured using various database tables, any of the removing, viewing, extracting and statistic operations require more complex SQL commands, which are implemented as specific stored procedures in GOP Tropo database. According to the relationships between

Table 3: List of other procedures for database content maintenance

Procedure	Arguments	Category	Remarks
vPoint	'source','site','description'	View	List filtered point list
vSource	'source'	View	List filtered source list
vSitel	'source','site','description'	View	List filtered site list
vData	'source','site','description'	View	List filtered data
vDiff	'source','site','description'	View	List filtered differences
fDataInfo		Info	Information on data table
fPairInfo		Info	Information on pair tables
fStatInfo		Info	Information on statistics tables
fDeletePoint	'source','site','description'	Delete	Remove selected data and site
fDeleteSite		Delete	Remove unreferenced sites
fDeletePair	'source','site','description'	Delete	Remove pairs and differences

records in various data tables, such procedures combine data for a transparent and user-friendly output. Additionally, for easy data selection, three common arguments for name or mask input (via a simple regular expression using asterix) are supported for most common maintenance procedures. Table 3 provides an incomplete list of maintenance procedures with respect to three category operations: view, delete or info.

While view procedures are designed to extract specific data combinations from various tables, the info procedures extract general information about table contents – e.g. start/end of data, number of records etc. All these procedures are, however, only minimum implementations to simplify common queries while any other specific query can still be requested using a standard SQL command.

On the contrary, delete procedures could be very tricky, in particular due to a possibility to lose a consistency within tables and their relations. Any removal is implicitly driven by the 'DELETE CASCADE' attribute used for most tables. The attribute specifies that when referenced row is deleted, row(s) referencing it should be automatically deleted as well. However, the cleaning of any specific point-/pair-related data or differences should be done with the relevant delete procedure. For cleaning sites in the *tSite* table which are not anymore referenced a specific stored procedure exists too.

Table 4: List of selected important settings from *postgresql.conf* configuration file

Name of variable	Value
shared_buffer	3000 MB
work_mem	512 MB
max_connections	10
maintenance_work_mem	256 MB
default_statistic_target	300
effective_cache_size	5000 MB
constraint_exclusion	partition

3. Database optimization

The database is running on a dedicated server with GNU Linux operating system Debian 6.0.7. and, currently, has reserved 12Gb memory and 8 thread 64-bit Intel(R) Xeon(R) CPU. This hardware configuration is sufficient for current operations. However, more than 750 millions of records are stored in each of the largest tables – tGNSS and tRAOBS – and there are more than one billion records total in the database, utilizing almost 100Gb of hard drive space. This amount of data can cause lack of performance and rapidly decrease execution speed of some SQL queries. The highest performance can be reached by optimal configuration of several parameters which are stored in *postgresql.conf* settings file, because they are set to extremely low values by default and, in most cases, they are not related to the available hardware configuration. For that reason, we discuss recommended and applied settings in the next paragraph.

The first important parameter – *work_mem* – defines the maximum limit of memory which can be used for one sorting operation. The amount of memory usage increases with each additional sorting. Each client connected to the same database server typically uses a maximum of two sorting operation at one time. This implies that the value of *work_mem* could be set to the amount of unused memory divided by the maximum number of connections while divided by two. The number of connections can be reduced in *max_connections* setting. When sorting, PostgreSQL estimates first the amount of memory for possible use. If *work_mem* value is not high enough, the system will use swap operations along with free space on the hard-drive, decreasing the performance. The parameter *shared_buffer* then determines the maximum amount of memory which can be used for cached data in memory after they are read from the hard drive. The higher value, the bigger the set of data is possible to store in memory, which reduces the number of swap operations. Optimal value is quite difficult to set, but 30% of the available memory is recommended for a dedicated server.

The VACUUM operation is one of the most important commands in PostgreSQL database server. It is a kind of garbage collector which releases allocated space from invalid records. The VACUUM operation is closely related to the ANALYZE operation which is used to collect statistics for optimizing a server query plan. The automatic maintenance of database is provided by the AUTOVACUUM command, which is turned on by default. The *maintenance_work_mem* parameter in the setting then defines the amount of memory which can be used for AUTOVACUUM. The last important parameter is *default_statistic_target*, value of which determines the quantity of data used for statistics. The higher the value, the higher the CPU utilization generated by server. On the other hand, a higher value of *default_statistic_target* likely yields more precise statistics and thus possibly increase the speed of the next SQL query. Table 4 summarizes the important setting of the PostgreSQL configuration file.

As it has been already mentioned above, the biggest table in GOP Tropo database contains more than half a billion of rows, which can increase easily when new data are introduced (e.g. such as from E-GVAP project). The SQL queries usually work with data in specific time range, for example, generating monthly statistics between different products. The PostgreSQL supports basic table partitioning while splitting single large table into several smaller pieces. This is done by applying an inherited scheme, which defines a parent table similarly as the original table, while data are stored in a sequence of child tables (partitions). Such partitioning can rapidly improve system performance, in particular when the query works with data in a single partition. On the other hand, additional overheads are relevant to all 'INSERT' commands calling a special trigger function when the table is partitioned. The inherited implementation guarantees that the partitioning has no direct influence to scripts or applications used within the database since data stored in child tables inherits behaviour from the empty parent table. In GOP Tropo database, range partitioning was applied in a yearly scheme on all large tables and each *tGNSS* child table thus contains about 20 millions records on average. New partitions are created by specific trigger PL/SQL functions called when data are inserted in the table. Table 5 summarizes the execution time of computing average value from data stored in a single partition restricted by WHERE clause. It is obvious that such query is much faster on the partitioned table than on the table without partitioning.

```
EXPLAIN ANALYZE SELECT AVG(ztd) from tGNSS as t1
where t1.epoch >= '2007-01-01 00:00:00' AND
           t1.epoch < '2008-01-01 00:01:00'
AND  t1.FK_Point = XX;
```

Table 5: The statistics from the partitioning test via analyze function for specific (repeated) command

Sequence number	Station	Source	Execution time (partitioned table)	Execution time (single table)
First	GOPE	EUREF Repro1	1363 [ms]	1300 [ms]
Second	GOPE	EUREF Repro1	13	48
Third	GOPE	EUREF Repro1	13	48
First	ALBH	IGS Repro1	47	125576
Second	ALBH	IGS Repro1	0.1	4422
Third	ALBH	IGS Repro1	0.1	4422

The database performance was tested on several different clusters as defined in Tab. 6. The original cluster is labelled as A and to this variant additional optimization steps were applied sequentially. In the first step the original file system (ext3) was replaced with newer file system (ext4) to improve swap operations (cluster B). The PostgreSQL setting was then revised according to that described above (cluster C). The *tGNSS* table was divided into yearly partitions (cluster D). Monthly partitioning produced a lot of tables and this variant was rejected from the comparisons. The *tPairDiff* table was also partitioned applying the yearly scheme (cluster E). The comparison averaging interval was decreased from 60 to 10 minutes (cluster F). The PostgreSQL was updated from 8.4 to recently newest 9.2 version (cluster G) with applying original settings and, finally, the settings was tuned also in PostgreSQL 9.2 in a similar way as for the version 8.4 (cluster H).

Table 6 summarizes the settings and performance of tested clusters. The statistics clearly show the importance of PostgreSQL configuration tuning, which improved overall performance by a factor of 2-4, in particular for all highly time-consuming procedures (e.g. inserting). That is true for the old as well as the new system version. The new file system also did not provide any improvements as expected, but even a slight degradation. In this general performance test, partitioning did not cause an increased performance when sets of procedures applying more complex SQL queries were used. This was not expected, but in general we decided to keep partitioning since only very small degradations and overhead costs were found. Further benefit can be expected for new PostgreSQL releases as well as in with growth of the data in database. The latter was the primary argument for preserving partitioned tables for large datasets.

Table 6: Variant settings of the database optimization and their performance

Cluster	File system	Partition data/diff	PostgreSQL vers/tuning	Compared samples	Insert [s]	Total [s]	Diff [s]
A	ext3	1/1	8.4 / no	60 min	4712	1130	835
B	ext4	1/1	8.4 / no	60 min	5050	1361	864
C	ext4	1/1	8.4 / yes	60 min	1348	944	660
D	ext4	Yearly/1	8.4 / yes	60 min	1663	1324	1033
E	ext4	Yearly / Yearly	8.4 / yes	60 min	1653	1377	1079
F	ext4	Yearly / Yearly	8.4 / yes	10 min	1637	7855	7563
G	ext4	Yearly / Yearly	9.2 / no	60 min	5759	1568	1228
H	ext4	Yearly / Yearly	9.2 / yes	60 min	1739	1554	1207

4. Evaluation examples

In order to demonstrate the initial database functionality, we provide several samples from routine evaluations with a focus on GNSS tropospheric products comparisons. The following products were used in the demonstration figures:

- IGS operational and re-processed (Repro1) tropospheric products (Byram, 2012),

- EUREF combined tropospheric results from the operational and re-processed (Repro1) solutions (Soehne and Weber, 2005),

- GOP global near real-time tropospheric product (Douša 2012b),

- GOP near real-time GPS and multi-GNSS(GPS+GLONASS) tropospheric products (Douša, 2012a).

We do not intend to go into details in the examples and we do not thus discuss details about product differences, e.g. applied software, processing strategy, models, constraints, precise products and many others affecting the ZTD solution. It is out of the scope of this paper to study effects of various changes clearly visible in statistics. Sample comparisons demostrate the calculated biases and standard deviations, either for the entire period or within period split into regular discrete intervals (e.g. years, months, weeks). The statistic results are provided with the database procedures for predefined configurations and the results are extracted from the database and visualized with Gnuplot (Williams and Kelley, 2011) or GMT (Wessel and Smith, 1998) plotting tools. Various figures are generated – total bias and standard deviations for the whole period and all common stations, the geographical distribution of the values or time-series of mean statistics over all stations from the comparison. The samples are given in Figs. 3-6.

Figure 3 shows the assessment of the GOP near real-time multi-GNSS ZTD solution with respect to the post-processed EUREF combined ZTD product over three months in 2011. For

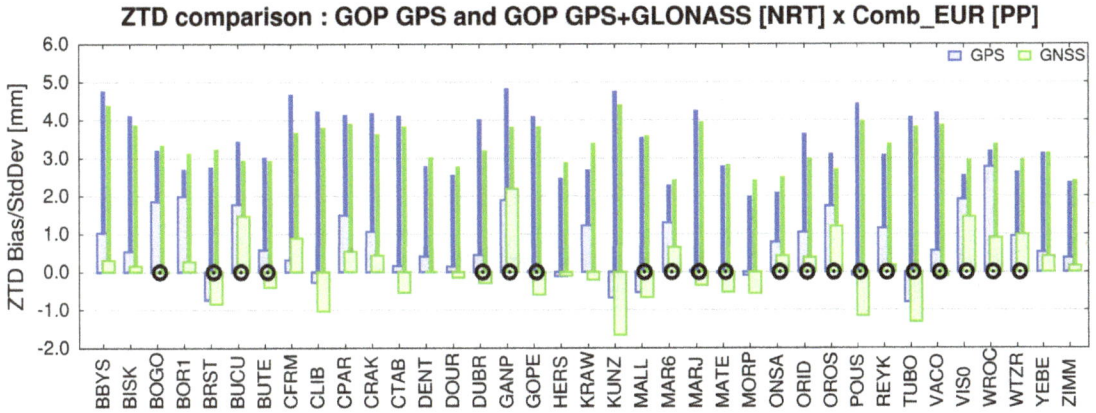

Figure 3: Example total bias and standard deviations for GOP GPS and GNSS ZTD near real-time solutions with respect to EUREF combined ZTD solution (three months in 2011). The circles indicates multi-GNSS stations.

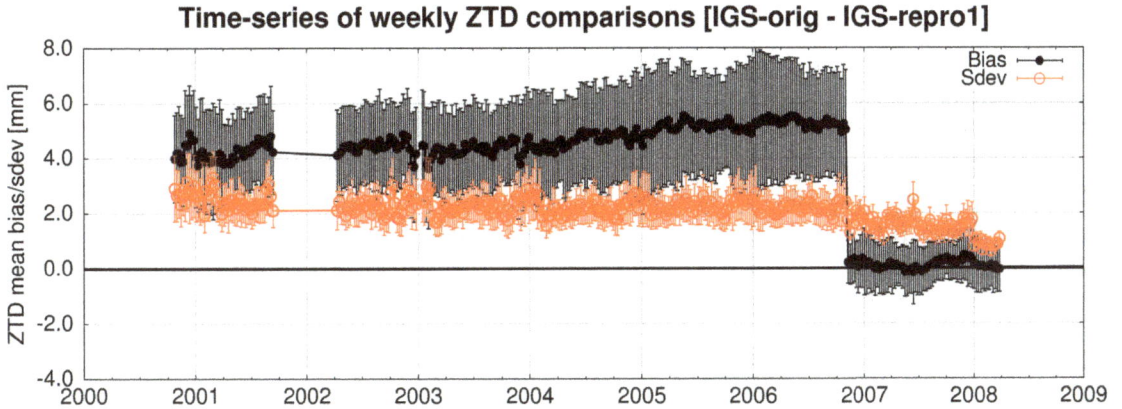

Figure 4: Weekly mean and its r.m.s. of ZTD biases and standard deviations from all stations

each station, which was identified as common to both products, the bias and standard deviation is calculated and plotted. Such plot provides information about the internal accuracy of GNSS ZTD products on a station by station basis. This is useful to assess a consistency of various strategies (and models) for comparison pairs when different product timeliness, input products, GPS or multi-GNSS observations and others are used.

Figure 4 and 5 show time-series of a long-term comparison of two products on a weekly and monthly intervals, respectively. The biases and standard deviations were calculated as mean values over all common stations for each interval individually so that any effect common to all stations can be visualized. Additionally, r.m.s. of such mean is calculated and plotted. Figure 4 compares historical and homogeneously reprocessed IGS ZTD products. The evolution of models and strategies within the historical product are clearly seen when compared to the re-processed ZTDs using the same strategy and models over entire period.

Figure 5 compares two different re-processing products – IGS (global) and EUREF (regional) while common stations in Europe could be compared only. Although both products are

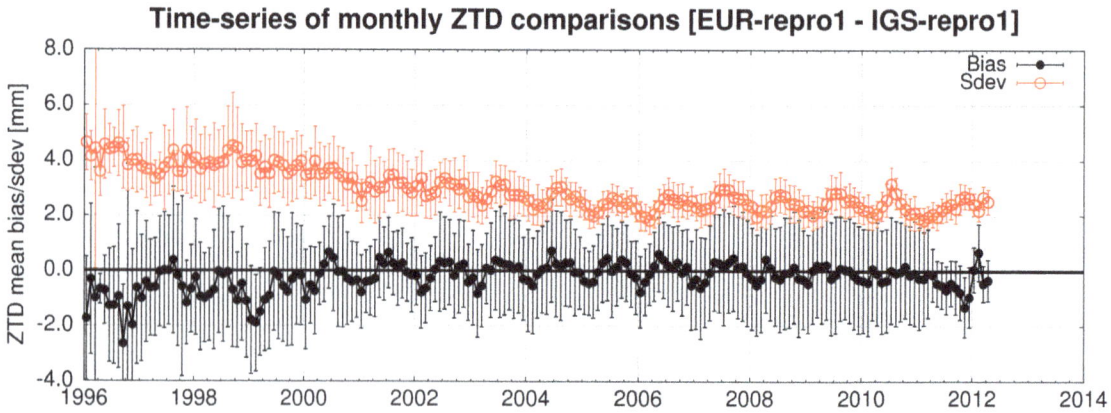

Figure 5: Monthly mean and its r.m.s. of ZTD biases and standard deviations from all stations

consistent over all time, the standard deviation clearly shows improvement in time, which can be attributed to the steadily increasing quality of data and products when more permanent stations in global and European networks are involved.

Finally, Figure 6 shows a comparison of IGS Final and IGS Repro 1 ZTD values for all dates covered by Repro 1. (Whereas Fig. 4 compared IGS Final and IGS Repro 1 ZTD estimates according to date, Fig. 6 compares them according to site.) The ZTD standard deviation is typically lower (about 2 mm) at low North hemisphere as well as in low latitudes in general. The effect of isolated stations, e.g. in central Asia, Africa and oceans, could be easily identified from the figure with standard deviations up to 4-5 mm. This can be attributed to the effect

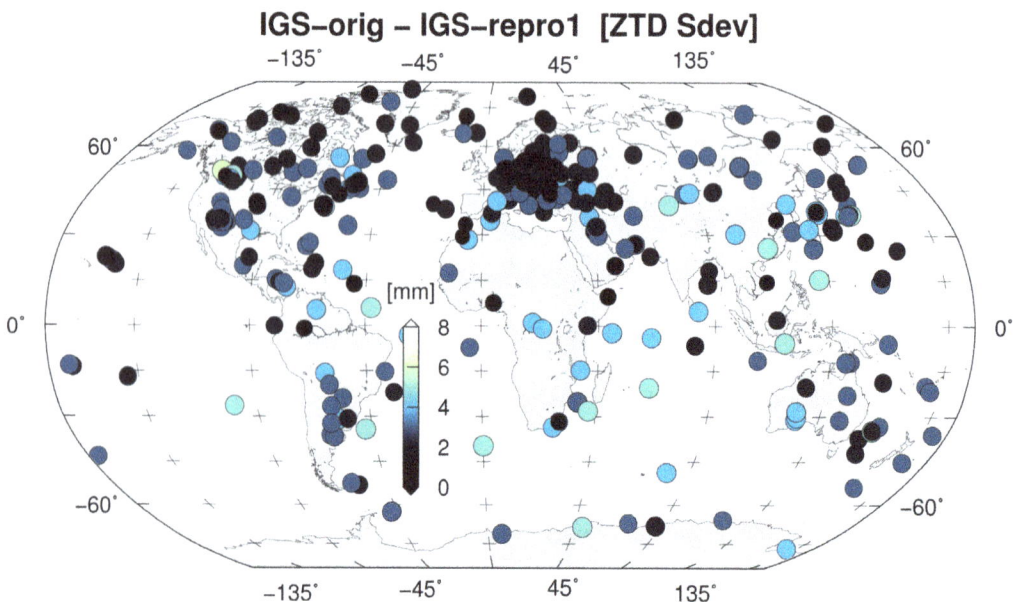

Figure 6: Displays geographical representation of ZTD standard deviations from two IGS global solutions

of lower accuracy of precise global orbit and clock products due to the lack of contributing stations, in particular during 90th.

5. Summary and outlook

The developments of the highly performance GOP Tropo database for the evaluation of tropospheric data and products were described with a special focus on its implementation aspects. The structure was designed in a flexible way to fulfil requirements specified in the introduction. Although the initial and primary motivation aimed for routine comparisons and evaluation of GNSS tropospheric products, current implementation functionality already goes beyond this scope. A special effort was given to the database optimization to support billions of records, which is already easy to achieve with several products. The optimization shows the need for revision of PostgreSQL default settings which could improve the overall performance by a factor of 2-4. Although careful partitioning did not improve performance, it was decided to keep it for the future since it is expected to become very important in handling huge quantities of data. Finally, samples of results were provided in order to demonstrate currently implemented functionality on selected interesting examples.

Acknowledgements

The database development was supported by the Czech Science Foundation (No. P209/12/2207). The authors also thank to Dr. Christine Hackman from U.S. Naval Observatory and two anonymous reviewers for useful comments improving the manuscript.

References

[1] Byram, S., Hackman, C. and Tracey, J. 2012. Computation of a High-Precision GPS-Based Troposphere Product by the USNO. Proceedings of the 24th International Technical Meeting of The Satellite Division of the Institute of Navigation (ION GNSS 2011). 2012, Portland, OR, September 2011, pp. 572-578.

[2] Douša, J. (2001): Towards an Operational Near-real Time Precipitable Water Vapor Estimation, Physics and Chemistry of the Earth, Part A, 26/3, pp. 189-194.

[3] Douša, J. (2003): Evaluation of tropospheric parameters estimated in various routine analyses, Physics and Chemistry of the Earth, 29/2-3, pp. 167-175.

[4] Douša, J. and G.V. Bennitt (2012): Estimation and evaluation of hourly updated global GPS Zenith Total Delays over ten months, GPS Solutions, Springer, Online-First.

[5] Douša, J. (2012): Development of the GLONASS ultra-rapid orbit determination at Geodetic Observatory Pecný, In: Geodesy of Planet Earth, S. Kenyon, M.C. Pacino, U. Marti (eds.), International Association of Geodesy Symposia, Vol 136, pp.1029-1036.

[6] Douša, J., Válavovic, P. and Gyori, G. 2013. Development of real-time GNSS ZTD products. presentation at the EGU 2013 General Assembly, April 7-12, 2013.

[7] Williams, T. and C. Kelley (2011). Gnuplot 4.5: an interactive plotting program. URL: http://www.gnuplot.info.

[8] POSTGRESQL, http://www.postgresql.org/docs/8.4/static/reference.html.

[9] Soehne, W. and G. Weber (2005): Status Report of the EPN Special Project "Troposphere Parameter Estimation", EUREF Publication No. 15, Mitteilungen des Bundesamtes fuer Kartographie und Geodaesie, Vienna, Austria, Band 38, pp. 79-82.

[10] Wessel, P. and W.H.F. Smith (1998): New improved version of the Generic Mapping Tools Released, EOS Tans. AGU, 79, 579.

## 9

Geostatistical Methods in R

Adéla Volfová, Martin Šmejkal
Students of Geoinformatics Programme
Faculty of Civil Engineering
Czech Technical University in Prague

Abstract

Geostatistics is a scientific field which provides methods for processing spatial data. In our project, geostatistics is used as a tool for describing spatial continuity and making predictions of some natural phenomena. An open source statistical project called R is used for all calculations. Listeners will be provided with a brief introduction to R and its geostatistical packages and basic principles of kriging and cokriging methods. Heavy mathematical background is omitted due to its complexity. In the second part of the presentation, several examples are shown of how to make a prediction in the whole area of interest where observations were made in just a few points. Results of these methods are compared.

Keywords: geostatistics, R, kriging, cokriging, spatial prediction, spatial data analysis

1. Introduction

Spatial data, also known as geospatial data or shortly geodata, carry information of a natural phenomenon including a location. This location allows us to georeference the described phenomenon to a region on the Earth. It is usually specified by coordinates such as longitude and latitude. By mapping spatial data, a data model is created. We will focus on a raster data model which provides value of the phenomena at each pixel of the area of interest. Mapping is a very common process in sciences such as geology, biology, and ecology. Geostatistics is a set of tools for predicting values in unsampled locations knowing spatial correlation between neighboring observations.

Making use of geostatistics requires difficult matrix computations briefly described in chapters Ordinary Kriging and Multivariate Geostatistics. In order to make our predictions easier, we are going to use methods from R geostatistical packages introduced in chapter R Basics. The best known geostatistical prediction methods are called kriging (for univariate data set) and cokriging (for multivariate data set) — examples of their use are shown in chapter Example of Kriging and Cokriging in R.

2. R Basics

There any many applications implementing geostatistical methods. Most of them are complex GIS and most of them are commercial. This does not hold for project R. R is a language and environment for any statistical computations and creating graphics. R is available as Free Software under the terms of the Free Software Foundation's GNU General Public License. R is multi–platform, easy to learn, and with a huge amount of additional packages that extend its functionality. For instance, such packages serve for special branches of statistics such as geostatistics.

2.1. First Steps in R

After downloading and installing R, open the default R console editor, figure 1. A standard prompt > appears at the last line. When this prompt is shown, R is ready to accept a command. Another prompt exists (+) which is for multiple row commands. There are two symbols for assigning a value to a variable: <- and =. When working in the *R Console*, the command is executed right after hitting *Enter*. If the user wishes to create a set of commands that are saved and can be used later, it is necessary to create a script (*File — New script*). For executing commands from a script, select all commands to be executed and press *Ctrl+R*. The results are shown in the *R Console*. The # symbol is used for comments. The following brief list of examples is the basic overview of commands to get you started in R. Please refer to countless internet tutorials for more advanced examples.

```
# Help and packages
help.start()                    # Load online HTML help
help(function)                  # Show online help for "function"
?function                       # Show online help for "function"
help.search("keyword")          # Open RGui dialog for packages/functions
                                # ... or classes connected to "keyword"
q()                             # Quit R
library()                       # List downloaded libraries
library(package)                # Load "package"
install.packages(package)       # Download "package"

# Assigning values
a <- 3
b = 8
x <- c(5, 2, 7)                 # Vector
y <- 2*x^2                      # Evaluate elements of vector x 1 by 1,
                                # ... dimension of x and y matches
y                               # Print content of a variable
    [1] 50  8 98                # ... result
1:5                             # Create sequence
    [1] 1 2 3 4 5               # ... result
seq(1,5)                        # Another way to create sequence
    [1] 1 2 3 4 5               # ... result
x[1]                            # Print 1st element of x
x[1:4]                          # Print 1st through 4th element of x
    5   2   7 NA                # ... result (when out of range,
                                # ... NA value is printed)
x[length(x)]                    # Print last element of vector

# Input table from a file
# ── file data.txt ──
Id   Price    Brand
1    3.5      Goldfish
```

```
2     7.0        Wolf
3     1.5        Seal
4     3.0        Goldfish
# ——————————————————

# Read table, a relative path can be used
# \ needs to be doubled
T = read.table("C:\\Data\\data.txt", header=T)

T                          # Print table T
T["Id"]                    # Print column Id
T[2]                       # Print 2nd column as data.frame
T[[2]]                     # Print 2nd column as vector
T[2,]                      # Print 2nd row
T[3,"Brand"]               # Print 3rd row in Brand column
T$Price                    # Print Price column into a row vector
colnames(T)                # Print column names
colnames(T) <- c(1,2,"x")  # Rename columns

# Plotting
x <- 1:5; y <- x^2
plot(x,y)                  # Plot data
points(x,y,pch="+")        # Add new data (use + symbols)
lines(x,y)                 # Add a solid line
text(10,12,"Some text")    # Write text on plot
abline(h=7)                # Add a horizontal line
abline(v=6,col="red")      # Add a vertical line
```

All necessary manuals and package documentation are stored in the Comprehensive R Archive Network (CRAN) [5].

2.2. Geostatistical Packages in R

A surprisingly large number of packages implementing geostatistical principles have been released. We are going to use only a few of them, the most common ones – geoR, gstat, and sp. However, for those who wish to explore more packages, a list of some spatial packages with a brief description is provided below. These packages were developed by different communities, therefore their functionality overlaps sometimes. Some of them are out of date and are not recommended for use anymore. Please refer to online documentation for more information on each package and its methods description.

geoR — is probably the most important package for geostatistical analysis and prediction.

geoRglm — extends functionality of geoR package. It is designed for dealing with generalized linear spatial models.

gstat — provides users with vast number of methods for both univariate and multivariate geostatistics, variogram modeling, and very useful plotting functions.

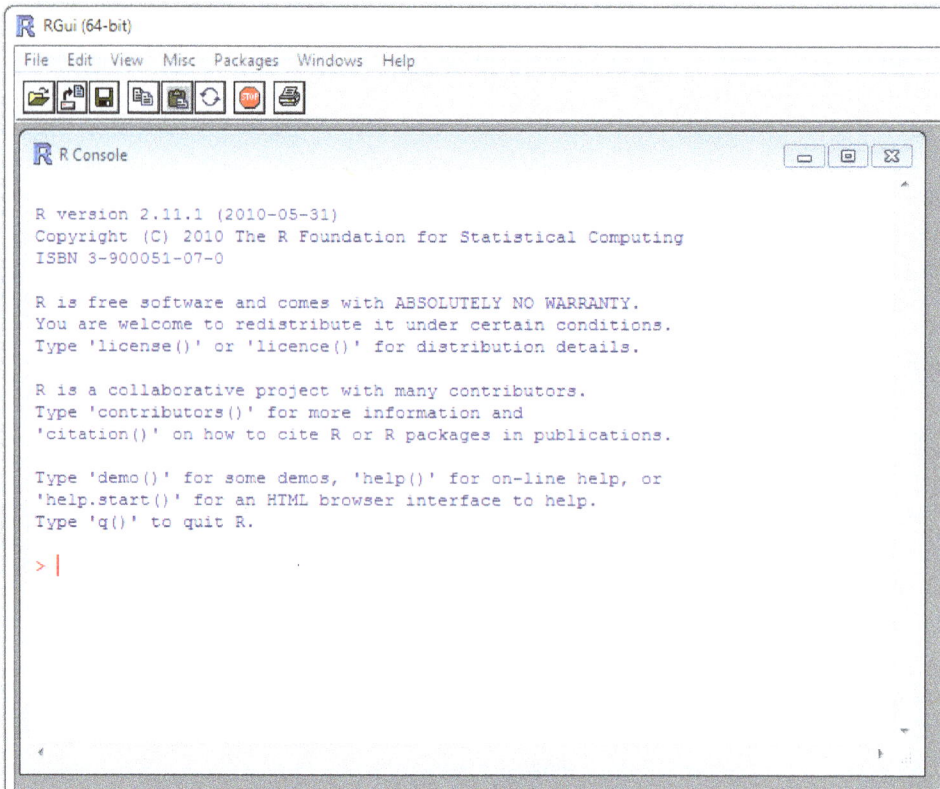

Figure 1: Default R interface.

sp — is a package for various work with spatial data — plotting, spatial selection, summary etc.. It also provides a very good training data set called meuse.

intamap — provides classes and methods for automated spatial interpolation.

fields — is a package with functionality similar to the gstat package. It is useful for curve, surface and function fitting, manipulating spatial data and spatial statistics. A covariance function implemented in R with the fields interface can be used for spatial prediction. This package also includes methods for visualization of spatial data.

RandomFields — provides methods for simulation and analysis of random spatial data sets. It also provides prediction methods such as kriging.

vardiag — allows to diagnose variogram interactivelly.

sgeostat — is an object–oriented framework for geostatistical modeling in S+[1].

spatial — contains methods for kriging and point pattern analysis.

spatstat — is another very extensive package for analysis of spatial data. Both 2D and 3D data sets can be processed. It contains over 1000 functions for plotting spatial data, exploratory data analysis, model–fitting, simulation, spatial sampling, model diagnostics, and

---

[1]S+ is a language for data analysis and statistics. It is possible to use the sgeostat package in R as well.

formal inference. Data types include point patterns, line segment patterns, spatial windows, pixel images, and tessellations.

There are many other packages dealing with spatial data in some way. A description of all of them is beyond the scope of this paper. For more information please refer to [4].

3. Spatial Statistics Basics

Spatial statistics is a set of statistical tools where location of data is considered. The main goal of geostatistics is to make a prediction of data x_i ($i = 1, 2, \ldots, n$) within an area of interest A where sample observations Z_i have been made. Each observation Z_i is dependent on values of a stochastic process $S(x)$ of spatial continuity in corresponding points x_i.

Functions in the `geoR` package are based on a Gaussian model. According to [8], chapter 3:

Gaussian stochastic processes are widely used in practice as models for geostatistical data. These models rarely have any physical justification. Rather, they are used as convenient empirical models which can capture a wide range of spatial behaviour according to the specification of their correlation structure.

Please, refer this book for more information on Gaussian processes.

3.1. Univariate Geostatistics

There are several statistics of spatial data that serve for general overview of the data set, show potential outliers among the observations, and describe the distribution. These features are shown in examples in section Analysis of Univariate Data.

Now, let's focus on the best known geostatistical method for prediction — kriging. There are many kinds of kriging. Each type determines a linear constraint on weights implied by the unbiasedness condition[2]. We are going to focus on ordinary kriging that assumes a constant but unknown mean.

In order to predict the phenomenon in the unsampled locations, we need to specify the spatial dependence. A geostatistical tool describing this dependence is called a variogram (figure 2). From now on, we assume isotropy in our data, then the variogram is so-called omnidirectional variogram. It is defined as the variance of the difference between field values at two locations across realizations of the field [6]. When shown as a plot, the x–axis represents the distance h between two observations. The maximum size h should be set such that we can expect two observations in this distance independent. The variance is depicted on the y–axis and is defined as:

$$\gamma(h) = \frac{1}{2n} \sum_{i=1}^{n} [Z(x_i) - Z(x_i + h)]^2,$$

where $Z(x_i)$ is an observed value of a random field and h is a distance between two observations. If the data are anisotropic, h becomes a vector and we need more variograms, each for a different angle. A variogram determined directly from the measurement is called empirical. For a prediction, we need to create a theoretical variogram that fits the empirical one as good as possible. The necessity of having a theoretical variogram lies in its continuity, so we can

---

[2]http://en.wikipedia.org/wiki/Kriging

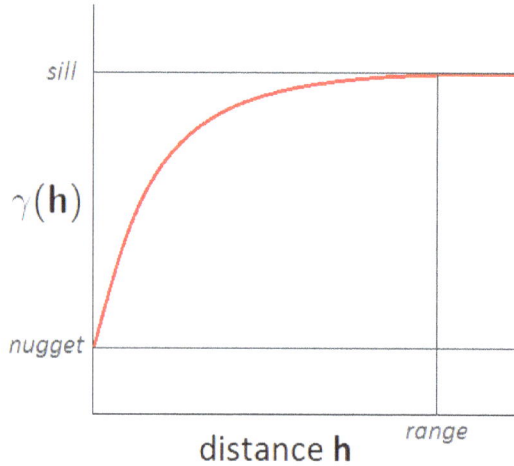

Figure 2: Variogram.

obtain the variance for any distance h. The kriging matrices based on such variogram must be positive definite.

There are three important characteristics of a variogram:

- *range* — a value of a variogram increases with increasing distance h up to a certain distance. Further than this, the variogram does not change much and we expect two observation independent behind this range,

- *sill* — the upper value of a variogram,

- *nugget* — the value of a variogram for zero h is strictly zero, nevertheless for the shortest distance h the variogram is computed, its value jumps from zero to a certain value (a nugget). This is called a nugget effect and it is caused, for instance, by an error of a measurement.

3.2. Ordinary Kriging

Since we have a model of spatial dependence (i.e. we know the formula of our theoretical variogram), we can predict the phenomenon in an unsampled location. Let us call this location x_0, then

$$Z^*(x_0) = \sum_{\alpha=1}^{n} \lambda_\alpha Z(x_\alpha),$$

where λ_α is a weight for value $Z(x_\alpha)$ at x_α.

Ordinary kriging is aliased BLUP (best linear unbiased predictor) and therefore the following conditions hold:

- a sum of weights is equal to 1 (guarantees the unbiasedness of the prediction),

- a variance of estimation errors is minimal.

There is one thing left to determine for the prediction — the vector of weights.

$$
\begin{bmatrix} \lambda_1 \\ \vdots \\ \lambda_n \\ \mu \end{bmatrix} = \begin{bmatrix} C_{11} & \cdots & C_{1n} & 1 \\ \vdots & \ddots & \vdots & \vdots \\ C_{n1} & \cdots & C_{nn} & 1 \\ 1 & \cdots & 1 & 0 \end{bmatrix}^{-1} \begin{bmatrix} C_{10} \\ \vdots \\ C_{n0} \\ 1 \end{bmatrix},
$$

where μ is a Lagrange parameter and C_{ij} is a covariance between $Z(x_i)$ and $Z(x_j)$. A relationship between a covariance and a variogram is following:

$$
C_{ij} = Cov(Z(x_i), Z(x_j)) = C(0) - \gamma(x_i - x_j),
$$

where $C(0)$ is the *sill* of the variogram model.

More detailed mathematical description is out of the scope of this paper. For more, please refer to [1,2,7].

3.3. Multivariate Geostatistics

Natural phenomena from one region can show some measure of dependency between each other. In such case, we can take one variable for prediction (primary) and the other variable(s) (secondary) to enhance the prediction. This is applied in cases where obtaining data of the primary variable is expensive, technically very difficult, or for any other reason we have an insufficient number of obtained data. In that case, we can look for some dependent variables in the region which we can measure in a much easier or cheaper way. Beside other advantages, we can reveal extreme values of the primary variable at locations where its measurement have not even been made.

3.4. Covariables Dependency

We assume to have only one secondary variable from now on. In case we have the covariables measured exactly at the same locations, we can easily tell the strength of their dependency by computing a correlation coefficient and/or by plotting a *scatterplot*. A *scatterplot* is a figure with axes corresponding to values of variables, one axis for each variable. In case we do not have measurement at the same locations, the best way to reveal the dependency is to compare variograms of the variables. We use so–called coregionalization when a cross–variogram is created [1].

3.5. Cross-variogram

A cross–variogram describes correlation between covariables and is given by:

$$
\gamma_{12}(h) = \frac{1}{2} E[(Z_1(x+h) - Z_1(x))(Z_2(x+h) - Z_2(x))],
$$

where Z_1 and Z_2 are primary and secondary variables. In some cases (e.g. in R methods), a pseudo cross–variogram is computed. There are inconsistent opinions on its use [1] (p. 150), however, its advantage is to gain much more points for an empirical cross–variogram. The pseudo cross–variogram is given by:

$$
\psi_{12}(h) = \frac{1}{2} E[(Z_1(x+h) - Z_2(x))^2].
$$

3.6. Ordinary Cokriging

Since we have the variogram and cross–variogram models, we can use ordinary cokriging for prediction. A value of a primary variable in an unsampled location is given by the following equation.

$$Z^*(x_0) = \sum_{S_1} \lambda_{1\alpha} Z_1(x_\alpha) + \sum_{S_2} \lambda_{2\alpha} Z_2(x_\alpha),$$

where S_1 and S_2 are sets of samples for the primary and secondary variables respectively.

The following hold:

- the sum of weights $\lambda_{1\alpha}$ is equal to 1 and the sum of weights $\lambda_{2\alpha}$ is equal to 0 (guarantees the unbiasedness of the prediction),

- a variance of estimation errors is minimal.

A relationship between a cross–variogram and a cross–covariance is:

$$\gamma_{12}(h) = C_{12}(0) - \frac{C_{12}(h) + C_{21}(h)}{2},$$

Then, ordinary cokriging system in matrix form is given as:

$$
\begin{bmatrix}
C_{11} & C_{12} & 1 & 0 \\
C_{21} & C_{22} & 0 & 1 \\
1 & 0 & 0 & 0 \\
0 & 1 & 0 & 0
\end{bmatrix}
\begin{bmatrix}
\lambda_1 \\
\lambda_2 \\
\mu_1 \\
\mu_2
\end{bmatrix}
=
\begin{bmatrix}
C_{01} \\
C_{02} \\
1 \\
0
\end{bmatrix},
$$

where C_{11} and C_{22} are covariance matrices of primary and secondary variables respectively, and C_{12} is a cross–covariance matrix.

For more detailed mathematical explanation of ordinary cokriging including proves, please follow [1,2].

3.7. Sampling Density and Location of Primary and Secondary Variables

There are several cases of how the covariables can be measured:

- samples of both, the primary and secondary variable, are obtained at exactly identical locations — this case is not very often because we either have a sufficient data set for the primary variable and so the secondary is not necessary to include in a prediction, or we do not have enough samples of the primary variable for creating a valid prediction by ordinary kriging and the secondary variable will not provide us with more useful information about it,

- the secondary variable is measured with higher density and all the primary variable samples overlap in location with the secondary variable — this is one of the most common cases of use of ordinary cokriging that leads to the best results; the primary variable measurement is not dense enough to make a good ordinary kriging prediction, so a significantly dependent secondary variable substantially increases the density of sampled locations and enhances the prediction of the primary variable,

- the secondary variable is measured with higher density and the covariables sample locations do not overlap — another very common case, however negatively effected by worse determination of the model of coregionalization which leads to not–so–good improvement of the prediction (compared to the previous case),

- number of samples of the secondary variable is smaller then number of samples of the primary variable — this case is useless for getting better prediction,

- the secondary variable samples correspond with all locations for prediction of the primary variable (very dense sampling) — for such a case another type of kriging is recommended — *kriging with external drift* [7]; the more dense the samples of the secondary variable, the harder the prediction to process when using ordinary cokriging.

See figure 3 for examples.

4. Spatial Statistics in R

For purposes of this chapter, sample data set from [2] has been used. These data are derived from a digital elevation model (DEM) of Walker Lake area (Nevada, USA) and are available in the `gstat` package, hence anyone can obtain the same data set a get to the same results.

4.1. Sample Data Set

The Walker Lake DEM has been modified for the sake of generality. There are 1.95 million points in the original data set. These points were divided into blocks of 5 by 5 points and final values were derived from them. There are two variables which we are going to use:

- U variance of the 25 values given by equation

$$U = \sigma^2 = \frac{1}{25} \sum_{i=1}^{25} (x_i - \bar{x})^2,$$

where $x_1, x_2, \ldots, x_{25}$ is elevation in meters,[3]

- V is function of mean and variance

$$V = [\bar{x} * \log(U + 1)]/10,$$

There are 78 000 values in a grid of 260 by 300 points. 470 points were chosen across the area to represent measurement.

4.2. Analysis of Univariate Data

A sample of 100 values in regular grid of 10 by 10 points is used for a following basic data description.

We can obtain a lot of useful information when we arrange the data according to some order, plot them or make some summary statistics. Beside other things, this is good for searching for outliers and errors in the measured data set.

---

[3]It is obvious that a flat terrain has low value of U, whereas hilly terrain has this variable very high. That is why this variable is also called as *topographic roughness index*.

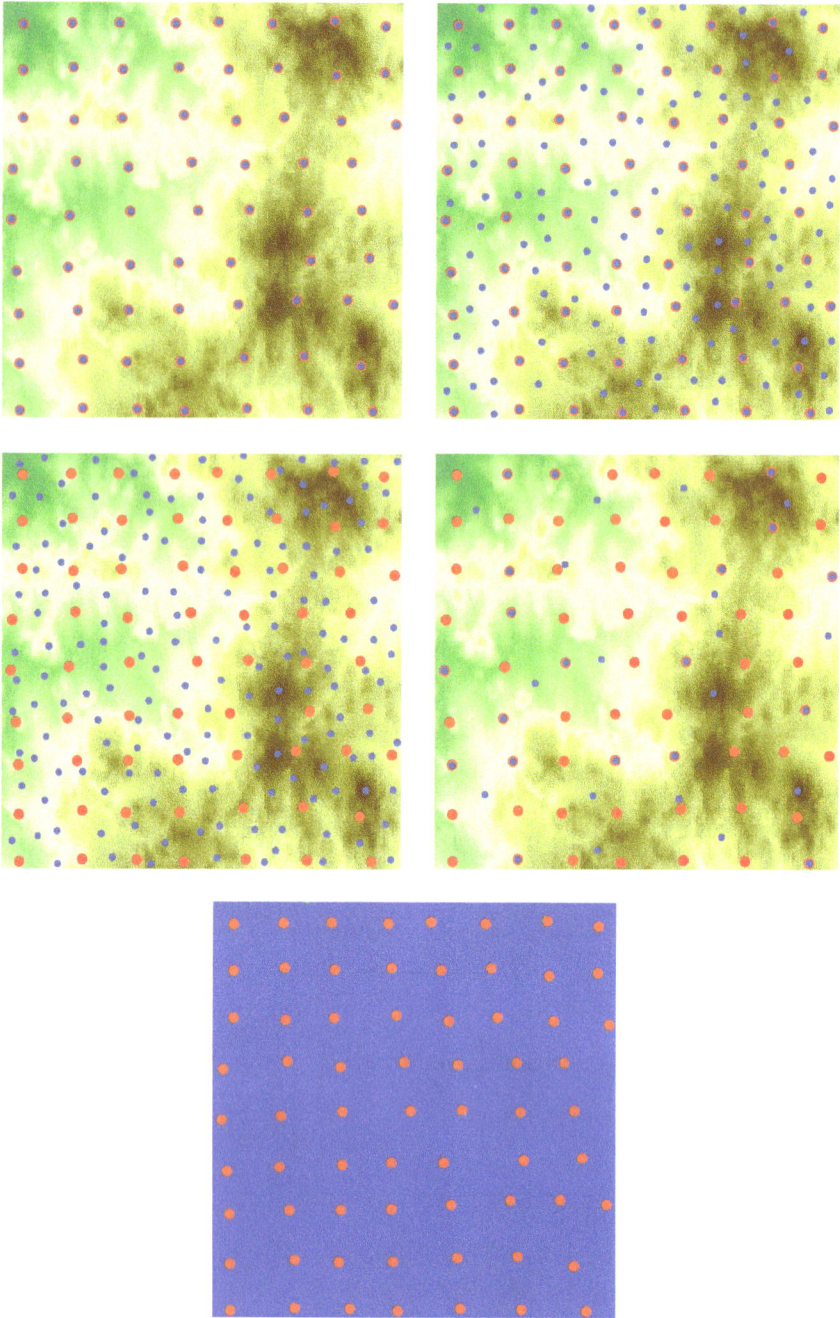

Figure 3: Location of primary and secondary variables. Order of the examples corresponds with chapter Sampling Density and Location of Primary and Secondary Variables, red dots stand for primary variable samples, blue represents secondary variable samples.

The most common presentation of the data is a frequency table and a histogram. The frequency table arranges the data into intervals and shows how many observations fall into each interval. An example of a frequency table for V data set is shown in table 1. See the datasets in the following listing:

U =

15	12	24	27	30	0	2	18	18	18
16	7	34	36	29	7	4	18	18	20
16	9	22	24	25	10	7	19	19	22
21	8	27	27	32	4	10	15	17	19
21	18	20	27	29	19	7	16	19	22
15	16	16	23	24	25	7	15	21	20
14	15	15	16	17	18	14	6	28	25
14	15	15	15	16	17	13	2	40	38
16	17	11	29	37	55	11	3	34	35
22	28	4	32	38	20	0	14	31	34

V =

81	77	103	112	123	19	40	111	114	120
82	61	110	121	119	77	52	111	117	124
82	74	97	105	112	91	73	115	118	129
88	70	103	111	122	64	84	105	113	123
89	88	94	110	116	108	73	107	118	127
77	82	86	101	109	113	79	102	120	121
74	80	85	90	97	101	96	72	128	130
75	80	83	87	94	99	95	48	139	145
77	84	74	108	121	143	91	52	136	144
82	100	47	111	124	109	0	98	134	144

A histogram o V values is shown in figure 4. An R function `hist` serves for plotting histogram.

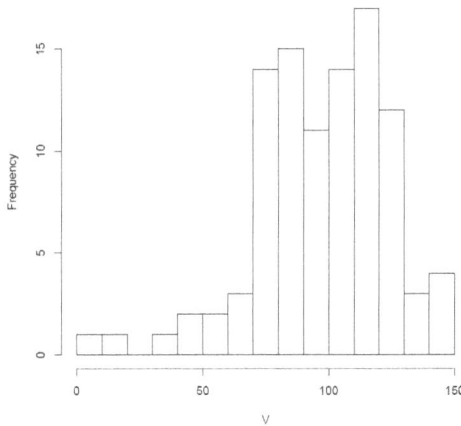

Figure 4: Histogram of V, `hist(V)`.

Some methods used later in this paper, works better for normally distributed data. We can tell whether the data are normally distributed from a plot with the measurement on the x–axis and cumulative frequency on the y–axis. In case of normal distribution, the points are arranged into a line. Example of this plot is in figure 5, it was created in R by calling `qqnorm`

Interval of V	Count
$(0, 10\rangle$	1
$(10, 20\rangle$	1
$(20, 30\rangle$	0
$(30, 40\rangle$	1
$(40, 50\rangle$	2
$(50, 60\rangle$	2
$(60, 70\rangle$	3
$(70, 80\rangle$	14
$(80, 90\rangle$	15
$(90, 100\rangle$	11
$(100, 110\rangle$	14
$(110, 120\rangle$	17
$(120, 130\rangle$	12
$(130, 140\rangle$	3
$(140, \infty\rangle$	4

Table 1: Frequency table for V.

function. Outliers and erroneous data, if some, can be observed in this plot.

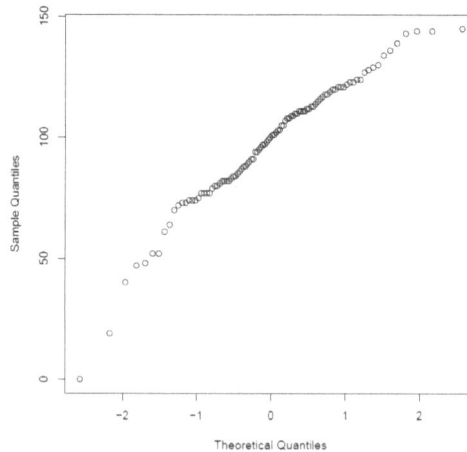

Figure 5: Visual test for normal distribution of V, `qqnorm(V)`.

Another plot good for visual exploration of the data is so–called box–and–whisker plot (figure 6). One half of the data lies inside the box. The line inside the box is median. The whisker lines represent a multiple of border values of the box (in this case a default 1.5 multiple was maintained). One can tell that values of $V(right)$ have larger variance in this case. The circles represent outliers. An R function `boxplot` has been used to create this plot.

For further data description, we look at the summary statistics such as the minimal and maximal value, mean, median, mode, quantiles, standard deviation, variance, interquartile range, coefficient of skewness, coefficient of variation etc.. First five mentioned statistics tell us about the location of important parts of the distribution. Next three values signify the variability of the distribution. The coefficient of skewness and coefficient of variation describe

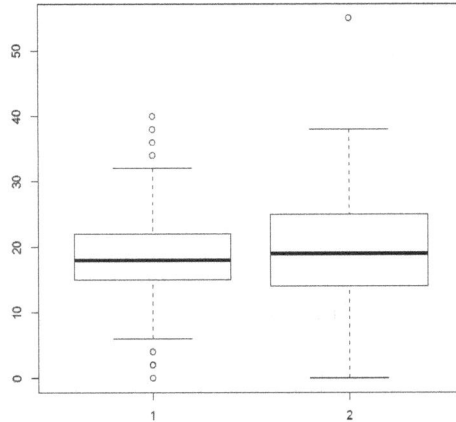

Figure 6: Box–and–whisker plot for U (left) and V (right), `boxplot(matrix(U,V))`.

the shape of the distribution and help to reveal potential erroneous observations.

We can obtain the basic statistics using the **summary** function. Summary statistics for V is listed in the following example.

```
> summary(V)
Min.    :   0.00
1st Qu.:  81.75
Median :100.50
Mean    :  97.50
3rd Qu.: 116.25
Max.    : 145.00

# variance
> var = sum((V-97.5)^2)/length(V)
689.69

# interquartile range
> IQR = 116.25-81.75
34.5

# coefficient of skewness
> CS = sum((V-97.5)^3)/sqrt(var)^3/length(V)
-0.771

# coefficient of variation
> CV = sqrt(var)/97.5
0.269
```

The coefficient of skewness is, in this case, negative which means the distribution rises slowly from the left and the median is greater then the mean. The closer the coefficient of skewness to zero, the more symmetrical the distribution. Hence the difference between median and mean is getting smaller.

The coefficient of variation is quite low. If this value is greater then 1, a search of erroneous observations is recommended.

4.3. Variogram

In order to plot an empirical variogram, we need to set a proper distance for the lag (x–axis on the plot). When the lag is too small, the variogram would go up and down despite its theoretical increasing trend before the range distance and constant trend for distance larger than the range. When we set the lag too large, we gain just a small number of values (breaks) on the variogram curve and we would not see the important characteristics of the variogram such as range, sill etc..

In our example we set the lag for 10 m. A variogram cloud (all pairs of points) and an empirical variogram with given lag for the V variable is in figure 7. These variograms were created by `variog` function. The theoretical variogram is modeled with `lines.variogram`

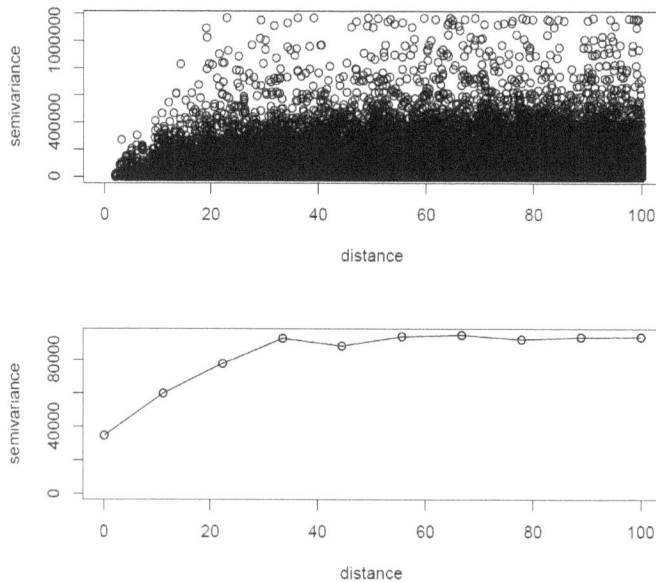

Figure 7: Variogram cloud and empirical variogram ($lag = 10\ m$) of V.

function or with an interactive tool `eyefit`. In our example in figure 8 we set the maximal distance to 100 m, the covariance model as exponential, the range to 25 m, the sill to 65000, and nugget to 34000.

4.4. Analysis of Multivariate Data

Since we wish to take advantage of spatial dependency of a primary and a secondary variable, we need to analyze the data sets. The goal is to examine whether the covariates are dependent enough so the secondary variable can improve prediction of the primary variable.

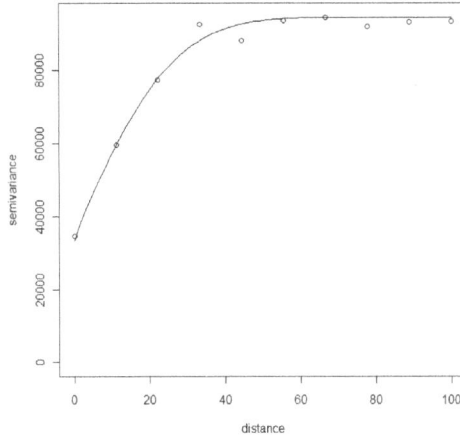

Figure 8: Theoretical variogram of V.

The first thing we can try is to compare the shape of histograms. Very similar shapes (i.e. similar distribution) indicates a certain degree of dependency.

By using `cor(U,V)` function in R we can get a correlation coefficient (in this case 0.837). Its value is always within the interval $\langle -1, 1\rangle$. The closer to zero, the less dependent the data sets are.

In order to compare two distributions, we can visualize so–called q–q plot (`qqplot` function in R). Each axis represents quantiles of one data set (see figure 9). If the plotted data are close to $y = x$ line, the variables are strongly dependent. If the data make a straight line that has a different direction than $y = x$, the variables still have similar distribution but with different mean and variance.

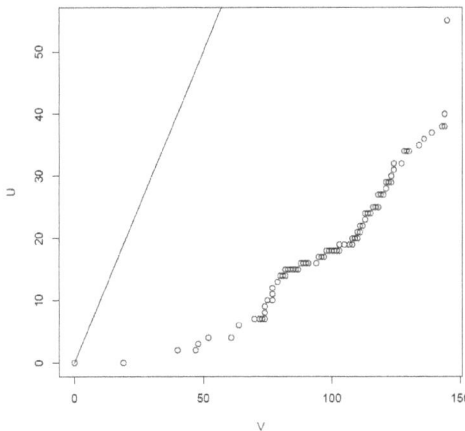

Figure 9: Q–q plot, straight line represents $y = x$, `qqplot(V,U)`.

Another graphical tool for testing the dependency of two spatial data sets is so–called *scatter-plot*. Pairs made of primary variable value and secondary variable value at the same location are visualized as points in this plot. The result is a cloud of points (see figure 10 for our example on U and V data). The narrower the cloud, the higher the degree of dependency. A *scatterplot* has one another big advantage — outliers and measurement errors lie outside the cloud. We can then easily check these points and in case they are wrong we would take them out of the data set. The dependency of two variables can be approximated by linear

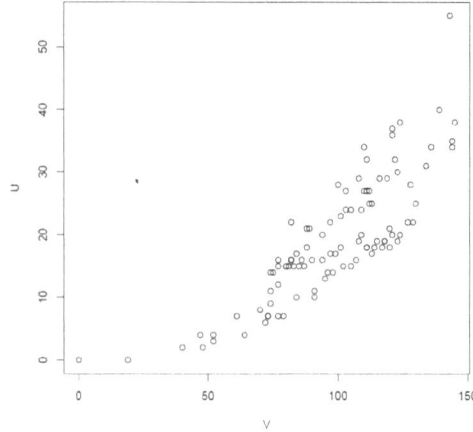

Figure 10: *Scatterplot*, `plot(V,U)`.

regression given by $y = ax + b$. How to do this in R is shown in the following code.

```
# method for linear regression
model = lm(U~V)

# plot
plot(V,U,main="Scatterplot and linear regression")
abline(model)

# model parameters
summary(model)
```

The plot from the previous example is in figure 11. An alternative for a linear regression can be a graph of conditional expectation where one variable is divided into classes (such as when we create a histogram) and a mean of the other variable is calculated within these classes, see figure 12.

Since we explored the data sets, did basic geostatistical analysis and determined the spatial continuity and covariables dependence, we might proceed to prediction. From now on, we are going to use a new data set that is more suitable as an example for prediction by (co)kriging.

4.5. Example of Kriging and Cokriging in R

In the following part of this paper, we are going to make two predictions — one using only primary variable on its own and ordinary kriging method, and the other using secondary variable and ordinary cokriging method. We are going to compare these two methods using some graphical and tabular outputs.

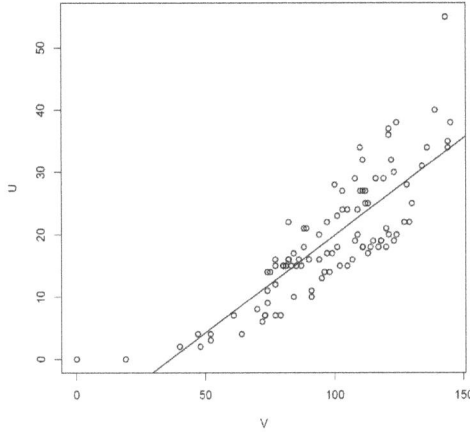

Figure 11: Linear regression $U = 0.314V - 11.5$.

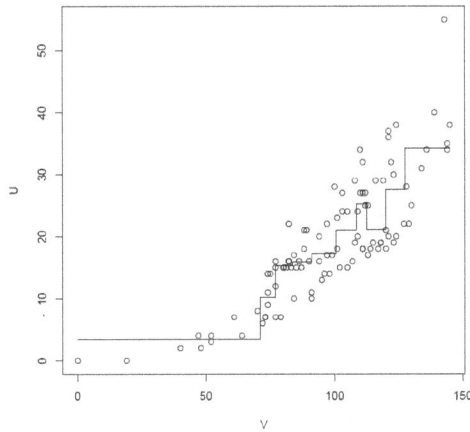

Figure 12: Conditional expectation of V within classes defined on U values.

4.6. Data Description

The phenomena we use in this example are simulated random fields in a square region of size of 50 pixels (i.e. 2500 pixels/values in total). We randomly[4] select some values and state them for measurement. After the prediction is made, we can easily compare the results with the original data set. This is not how it works in reality — we do not have values of the variable at each location of the region, that is why we do the prediction. However, for educational purposes, comparison of predicted and real values is a good way to show how these methods work and how well they work.

Simulation of Gauss Random Fields was chosen to create our phenomena by method `grf` in R. This method is able to create a random raster which can represent continuous spatial phenomenon. Gaussianity of the spatial random process is an assumption common for most standard applications in geostatistics. However non-Gaussian data are often provided. How

---

[4]The layout of the samples is not random — we try to cover the whole region and arrange the samples in a grid. However, the samples are randomly chosen from a neighborhood of each node of the grid.

to deal with this sort of data is decribed in detail in [9].

In our paper, two fields were created by function `grf`, each representing one variable (called A and B). A is our primary variable for which the prediction will be made. B is just an auxiliary variable for forming the secondary variable – C. C is strongly correlated with A, the correlation coefficient is about 0.93. All three fields are shown in figure 13. The R code of creating and plotting these three fields is following:

```
library(geoR)
library(gstat)

set.seed(1)
# creates regular grid of 50 by 50 pixels
# the covariance parameters are sigma^2 (partial sill)
# and phi (range parameter)
A = grf(50^2,grid="reg",cov.pars=c(1,0.25))
# all values of A are non-negative
A$data = (A$data+abs(min(A$data)))*100

set.seed(1)
# covariance model is set to matern
# smoothness parameter kappa is set 2.5
B = grf(50^2,grid="reg",cov.pars=c(1200,0.1),
    cov.model="mat",kappa=2.5)

C = A
C$data = A$data-B$data
# all values of C are non-negative
C$data = C$data+abs(min(C$data))

library(fields)
img_A = xyz2img(data.frame(A))
img_B = xyz2img(data.frame(B))
img_C = xyz2img(data.frame(C))

par(mfrow=c(2,2))
image.plot(img_A, col=terrain.colors(64), main="A",
           asp=1, bty="n", xlab="", ylab="")
image.plot(img_B, col=terrain.colors(64), main="B",
           asp=1, bty="n", xlab="", ylab="")
image.plot(img_C, col=terrain.colors(64), main="C",
           asp=1, bty="n", xlab="", ylab="")
```

Both, A and C, have normal distribution, and all values are non–negative for sake of easier presentation. The coordinates are in range $\langle 0, 1 \rangle$. The basic statistics are in table 2.

The sample data set consist of 166 measured values of C and 63 values of A. The primary variable fully overlaps the samples of the secondary variable and the secondary variable sample

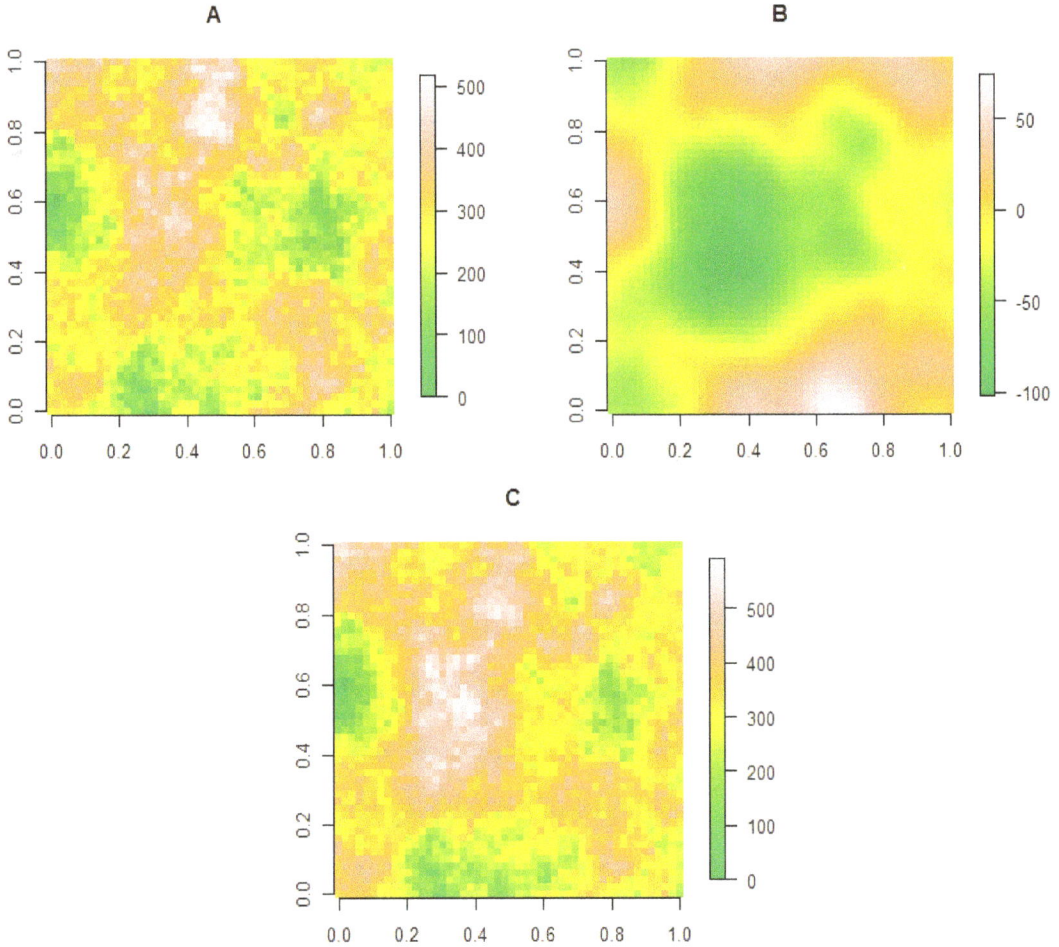

Figure 13: Simulated random variables.

grid is much more dense (see later in figure 18). Let us have a look at some analyzing graphical tools — histograms of samples are shown in figure 14, q–q plots are shown in figure 15, and a *scatterplot* is shown in figure 16. According to these plots, we can conclude that the samples have normal distribution and the distributions are quite similar which confirms the strong correlation of the variables.

4.7. Prediction Using Ordinary Kriging

Use of ordinary kriging in R is very simple. Once we determined the theoretical variogram we can proceed to the prediction. See the following code:

```
# create a grid for the prediction
gr = data.frame(Coord1=A$coords[,"x"],Coord2=A$coords[,"y"])
gridded(gr) = ~Coord1+Coord2

# assign coordinates to variable A
```

Variable	Number of values	Minimum	Median	Mean	Maximum
A (primary)	2500	0.0	289.8	286.1	517.0
C (secondary)	2500	0.0	342.8	342.6	590.5

Table 2: Basic statistics of primary and secondary variable.

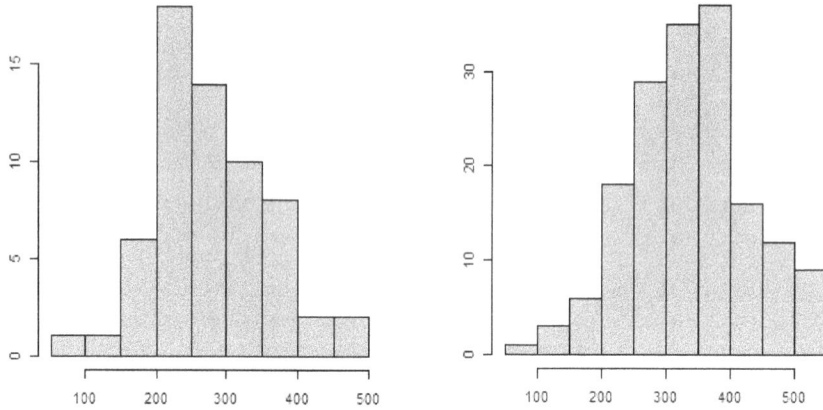

Figure 14: Histograms of A (left) and C (right).

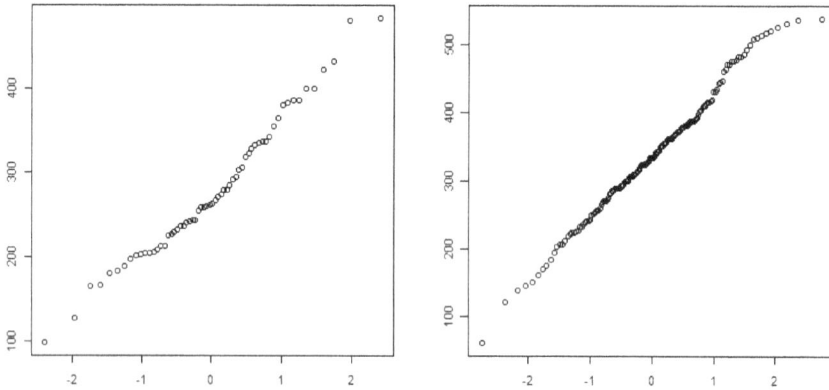

Figure 15: Q–q plots of A (left) and C (right).

```
coordinates(dataFrameA) = ~Coord1+Coord2

# variogram model
vm = variogram(data~1,dataFrameA)
vm.fit = fit.variogram(vm, vgm(6500, "Sph", 0.3, 50))

# prediction using ordinary kriging
OK_A = krige(data~1,dataFrameA,gr,vm.fit)
```

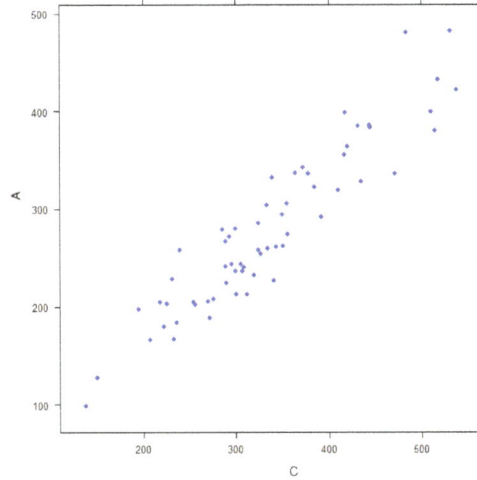

Figure 16: *Scatterplot* of *A* a *C*.

This is all we need to do to get prediction in unsampled locations when input is only the primary variable *A*. The results are shown in figures 18 and 19. Let us have a look how the process changes when we wish to include the secondary variable.

4.8. Prediction Using Ordinary Cokriging

A detailed description of how to process ordinary cokriging prediction in R is decribed in [3].

We already concluded that the variables *A* and *C* are spatially dependent. The most difficult step in prediction by ordinary cokriging is to set a linear model of coregionalization (in other words, to describe the spatial dependence between the covariables). We need to fit the samples into proper variogram and cross–variogram models. Follow the example in the code below:

```
# create a gstat object g
# (necessary for correct use in following methods)
# varibles A and C are saved in class data.frame
# add A and C to object g
g <- gstat(NULL, id = "A", form = data ~ 1, data=dataFrameA)
g <- gstat(g, id = "C", form = data ~ 1, data=dataFrameC)

# empirical variogram and cross-variogram
v.cross <- variogram(g)
plot(v.cross, pl=T)

# add variogram to object g
# vmA_fit is previously created variogram model
g <- gstat(g, id = "A", model = vmA_fit, fill.all=T)

#create linear model of coregionalization
g <- fit.lmc(v.cross, g)
```

```
plot(variogram(g), model=g$model)
```

The model of coregionalization is shown in figure 17. The upper figure is variogram of samples of A. The empirical variogram does not look good due to small number of input samples. Look at the improvement of variogram for C (lower right) where the number of samples is about three times larger. The lower left figure is the pseudo cross–variogram. The covariance model is identical (spherical in this case) for all three variograms, as well as the range was maintained (about 0.3). This means that the covariables behave similarly in space — they show the same degree of dependence for given distance. Since we gained linear model of coregionalization, we can proceed to prediction using ordinary cokriging.

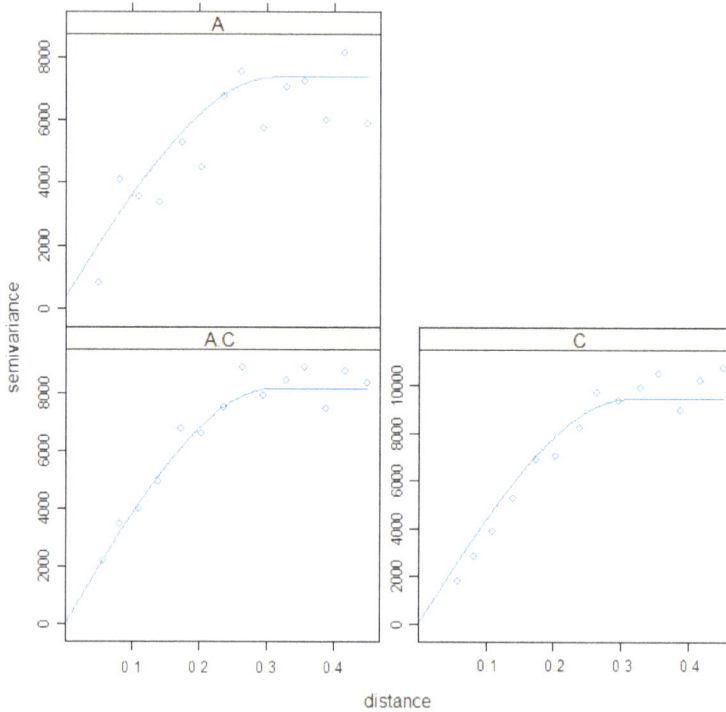

Figure 17: Variogram and pseudo cross–variogram of A and C.

The prediction step in R is actually very simple. It is literally a single command of method `predict.gstat` method. This method distinguishes (based on input data) what prediction method to use. There are actually two predictions made. One for our primary variable and one for the secondary one, because the method does not make a difference between those variables (i.e. we never specify which one is the primary one).

```
# gr is the prediction grid
CK <- predict.gstat(g, gr)
```

Comparisons of some statistics are listed in table 3. The contribution of C variable to the prediction of A is obvious. The extreme values got closer to real extreme values of A. The same holds for the median and mean. Values of variation of prediction got significantly lower.

Data	Min.	Med.	Mean	Max.	Mean of var.pred.	Max.var. of pred.
A real	0.0	289.8	286.1	517.0	–	–
OK A	98.2	277.4	280.8	482.4	2804	5329
CK A, C	42.3	279.6	281.0	482.4	1617	3293

Data	Min. diff.	Mean diff.	Max. diff.	Med. of diff.	Med. of abs(diff.)	RMSE
OK A	-153.5	-5.1	179.1	-4.1	32.2	49.1
CK A, C	-177.0	-5.1	166.4	-3.0	30.8	46.8

Table 3: Comparison of ordinary kriging (OK) a ordinary cokriging (CK).

RMSE stands for root mean square error:

$$RMSE = \sqrt{\frac{1}{n} \sum_{i=1}^{n} (Z^*(x_i) - Z(x_i))^2}.$$

The best result presentation is visualization of the predictions (figure 18) and the prediction errors (figure 19). It is obvious that the cokriging prediction describes the regions with extreme values more precisely. However, we can see that the kriging prediction did a good job too. It is thanks to relatively sufficient number of samples and (more importantly) their proper layout. It is only on us to decide whether this prediction is accurate enough or not. If not, we need to provide the prediction with samples of another variable that is highly correlated with the primary one and that has more dense sampling. The question is whether the improvement is worth the cost of the secondary variable data set. Let us pay attention to the errors figure, particularly on the middle map with real errors. We can see that in case of ordinary cokriging a red cloud of errors appeared in the middle. This is a somewhat negative impact of the C samples. Let us recall that the C variable is derived not only from A but also from B variable (figure 13) that has a large region of negative values exactly in the place where the red cloud of errors appeared. This region effected the C samples as well as the final prediction of A. This may have a dangerous impact on the prediction when using a secondary variable. This is why the degree of dependency of the covariables has to be really high.

5. Conclusion

Both methods, ordinary kriging and ordinary cokriging, were shown to lead to a successful prediction. As we expected, the gain of the secondary variable was obvious. However, we always need to consider the cost of obtaining it and a the quality of the prediction without it. We did much more combinations of covariables during this project that were not mentioned in the paper. We worked with yet another variable that was not so correlated to the primary one. The results in that case were not good which we expected. We tried different sample layouts for primary and secondary variable. The biggest gain in prediction was achieved when the primary data set was so sparse that prediction by ordinary kriging was almost impossible to process (we cannot create the variogram). By adding the secondary variable, the prediction gave us quite decent results. We also tried to use the same primary variable as in this paper and the secondary variable just with the difference in sample locations — they did not overlap with the primary variable samples (their count was still about three

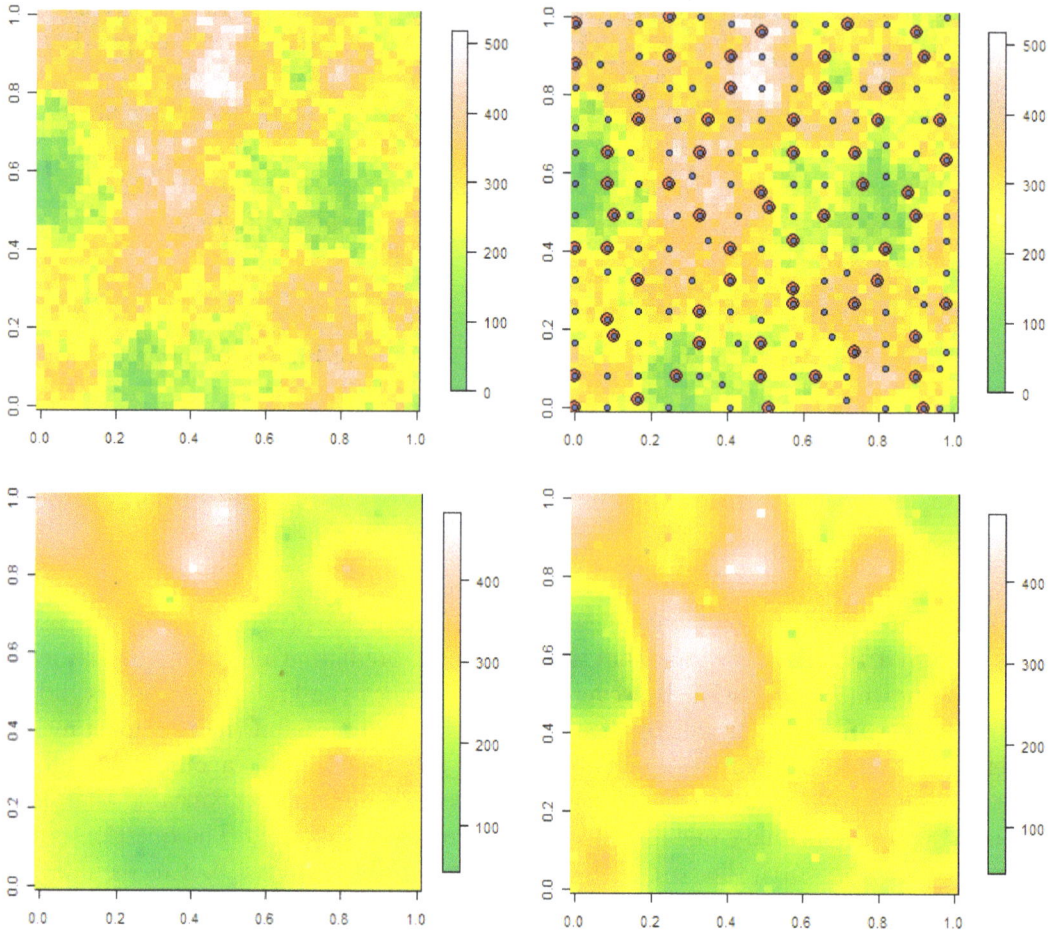

Figure 18: Ordinary kriging and ordinary cokriging for A and C (left upper – real values of A, right upper – samples (red – A, blue – C), left lower – ordinary kriging, right lower – ordinary cokriging).

times higher than number of samples for primary variable). This is the case where we cannot tell how good the spatial dependency of the covariables is and so it is harder to create the linear model of coregionalization. Results of such prediction were not that good as in the case presented in this paper, however we still managed to enhance the prediction of the primary variable.

This paper was originally made for educational purposes. It shows how to do basic spatial data analysis and how to predict values of some phenomenon in unsampled locations. Two methods were described — ordinary kriging and ordinary cokriging. Readers of this paper were provided with a step–by–step prediction process in R environment.

Acknowledgments *The project was supported by grant SGS11/003/OHK1/1T/11. Many thanks belong to Prof. Dr. Jürgen Pilz who became a great inspiration leading to including geostatistics and project R into the Geoinformatics programme at the Czech Technical University.*

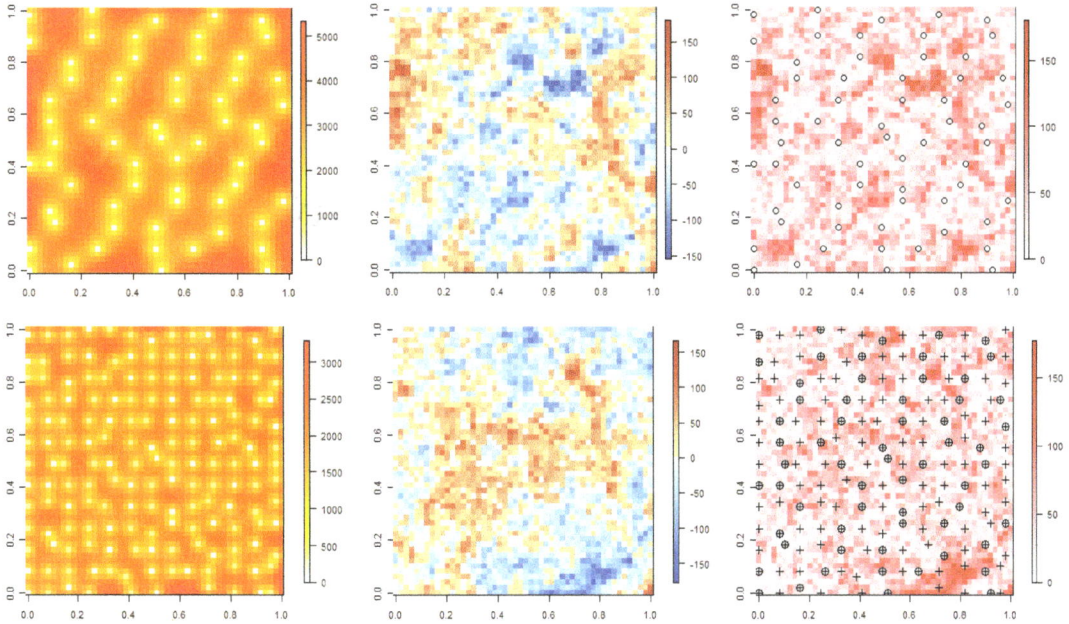

Figure 19: Prediction errors. Upper row for ordinary kriging, lower row for ordinary cokriging; Left: Variation of prediction, middle: Real estimation errors, right: Absolute values of estimation errors (circle – A, plus – C).

References

[1] Wackernagel, H. (2003): Multivariate Geostatistics. - 3rd edition. - Springer, Germany.

[2] Isaaks, E. H.; Srivastava, R. M. (1989): Applied Geostatistics. - Oxford University Press, New York.

[3] Rossiter, D. G.: Co-kriging with the gstat Package of the R Environment for Statistical Computing. - Web: http://www.itc.nl/ rossiter/teach/R/R ck.pdf.

[4] CRAN Task View: Analysis of Spatial Data. - Web: http://cran.r-project.org/web/ views/Spatial.html.

[5] The Comprehensive R Archive Network. - Web: http://cran.r-project.org.

[6] Cressie, N. (1993): Statistics for spatial data. - Wiley Interscience.

[7] Hengl, T.: A Practical Guide to Geostatistical Mapping. - 2nd edition. - Office for Official Publications of the European Communities, Luxembourg. - Web: http://spatial-analyst.net/book/.

[8] Diggle, P. J.; Riberio, P. J. Jr. (2007): Model–based Geostatistics. - Springer.

[9] Pilz, J. (Ed.) (2009): Interfacing Geostatistics and GIS. - Paper: Bayesian Trans-Gaussian Kriging with Log-Log Transformed Skew Data by Spöck G., Kazianka H., and Pilz J.. - Springer.

## 10

Results of GPS reprocessing campaign (1996-2011) provided by Geodetic observatory Pecný

Jan Douša, Pavel Václavovic

Research Institute of Geodesy, Topography and Cartography, Geodetic Observatory Pecný Ústecká 98, Zdiby 250 66

jan.dousa@pecny.cz

Abstract

The paper presents the GOP first reprocessing results, which officially contributed to the EPN-repro1 project. It also describes the 15-year GOP cumulative solution providing station coordinates, velocities and their discontinuities over the period of 1996-2011. Repeatabilities estimated from cleaned long-term coordinate time-series reached 1-2 mm and 4-6 mm in horizontal and vertical component, respectively. We then showed the exploitation of GOP reprocessing results in the assessment of the EUREF ITRF2005 densification and the latest ITRS realization, ITRF2008. We identified and confirmed the North-South tilt ($\approx 2mas$) in the currently available European reference frame based on the EPN cumulative solution updated in GPS week 1600. The study showed a historical development of the tilt and its close relation to a weak velocity datum definition of this realization, which is very important for a long-term datum prediction. Selected EPN station coordinates, velocities and discontinuities of the latest ITRS realization (ITRF2008) were also assessed. Specific problems for some EPN stations were identified in the global reference frame. This emphasized further necessity to check all the stations before their use for datum definition for regional densifications.

Keywords: Global Positioning System, permanent network, reprocessing, terrestrial reference frame, cumulative solution, coordinates, velocities, discontinuities

1. Introduction

The EUREF (EUropan REference Frame, http://www.euref-iag.org) is one of the subcommisions for regional reference frames of the International Association of Geodesy (IAG). The EUREF's main goal is to define, to realize and to maintain terrestrial and vertical reference systems in Europe. To serve both scientific and practical applications in positioning and navigation, the European terrestrial reference system (ETRS89) has been realized mainly using GPS observations as they became available at the beging of 90th. The EUREF permanent network (EPN, e.g. [1]) was proposed by W. Gurtner in 1995 [2] in order to provide a continuously operating ground-based infrastructure for such realization of ETRS89. Since that time EPN has developed from 40 stations (1996) to about 300 stations today (September, 2012), almost regularly distributed over Europe and the surrounding areas.

In order to handle a unique and common solution for all EPN stations, a distributed processing scheme of GPS data was organized from the beginning. EUREF subnetworks, routinely processed by 17 local analysis centers (LACs) on a weekly basis, contribute to an EPN weekly combined product. European terrestrial reference frame (ETRF), which represents a realization of ETRS89, is based on a long-term combination of EPN weekly solutions. Such combination includes the estimation of station coordinate changes per year (later referred simply as 'velocities') in addition to reference coordinates at a central epoch. Together with a global terrestrial reference frame (labeled as ITRF<year>), which is maintained by the International Earth Rotation and Reference Systems Service, IERS [3], European reference frame (ETRF<year>) is established based on densification solutions provided by EPN.

During years, however, the GNSS data analysis of global (and regional) networks were affected by different factors due the updates of reference system realizations, processing models, analyses strategies and software packages. Precise products provided by the International GNSS Services (IGS, [4]) – orbits, Earth Rotation parameters and clock corrections – are dependent on the factors available at the time of generating these products. Thus the global products are not consistent over the past period (1996-2012) and, consequently, also the coordinates estimated by the EPN LACs. Inconsistencies in the time series of the coordinates appear whenever reference frame or commonly adopted models were changed. Global reprocessing activities were started by the Potsdam-Dresden Group [5] and later provided also by other analysis centers in IGS-repro1 project [6] coordinated by IGS. The global reprocessing activities included the generation of improved global products labeled as IGS-repro1 products.

In the meantime some LACs of the EPN also started regional reprocessing activities in order to provide initial solutions homogeneous over a whole period, i.e. without changes in reference frames, global products, processing strategy or applied models. A plan of the coordinated reprocessing activity, spanning over period 1996–2009, was agreed at the EUREF LAC workshop in October 2008. The redistribution of EPN stations was organized (not all LACs were able to contribute), historical archive established, benchmark campaign prepared and evaluated and processing strategies agreed. The solutions of individual subnetworks were then finished in 2011 and combined EPN weekly solutions prepared. The densification of up-to-date global terrestrial reference frame (ITRF2008) could be thus based on a long-term combination of all European stations.

The Geodetic observatory Pecný (GOP) has contributed to EPN as a EUREF local analysis center since January 1997, see [7]. In 2010-2011, GOP LAC processed an extended subnetwork for the EPN reprocessing project. This paper describes daily/weekly GOP solutions and a long-term cumulative (Sec. 2). The GOP-repro1 results were then exploited in the assessment of European ITRF2005 densification based on the EPN cumulative solutions (Sec. 3) and in the assessment of station discontinuities available from the latest terrestrial reference system realization, ITRF2008 (Sec. 4). Section 5 sums up with concluding remarks.

2. GOP-repro1 solution

This section gives a brief summary of the GOP-repro1 solution which was prepared using the Bernese GNSS software V50 [8] and an in-house developed processing engine. At the first level the processing was carried out by analysing daily data batches and followed with combinations of daily solutions into weekly ones. Figure 1 shows the EPN stations processed by GOP for the EPN-repro1 project. The original subnetwork (stations marked in black) was extended to support the EPN-repro1 requirements that each station is processed by three analysis centres at least.

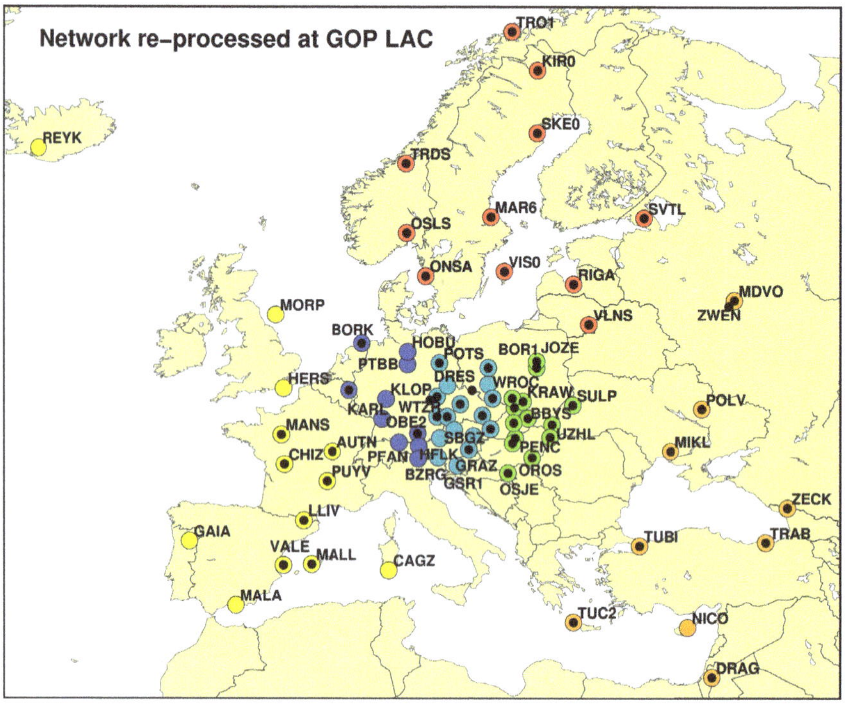

Figure 1: Map of GOP subnetwork extended for the EPN-repro1 project. Different colors shows processing clusters and black color shows the stations from the routine GOP contribution to EPN.

2.1. Processing strategy

The global IGS-repro1 products (orbits and Earth rotation parameters) were estimated using the IERS 2003 Conventions [9]. Fixing these global products in regional daily reprocessing, consistent conventions and models needed to be applied, which was followed in most cases, but these two: (1) models were not supported by the latest release of the software package (e.g. the newest tropospheric models in the Bernese GNSS software V5.0), (2) individual antenna calibration models used in EPN, while the IGS follows type-specific models only.

The daily processing consists of the following steps – converting input data and products; preparing a priori coordinates and information; synchronizing receiver clocks; generating single differences over baselines while keeping a maximum number of observations; cleaning data, detecting cycle slips (based on triple-difference approach) and setting ambiguities; screening post-fit residuals and detecting outliers; selecting fiducial stations for datum definition; resolving integer ambiguities; generating ambiguity-fixed daily solutions. The weekly products then include combinations of corresponding daily solutions based on a normal equation stacking procedure. The tropospheric products are finally generated for each day of a week based on a fixing weekly coordinates in new daily adjustments.

The success of resolved integer ambiguities using the QIF strategy [10] is shown in Figure 2. It is strongly seasonally dependent, however, a significant decrease of the percentage of fixed ambiguities is also visible during 1997-2003 which could be partly attributed to the high solar activity (with maximum in 2000) and partly to the lower quality of GPS observations and

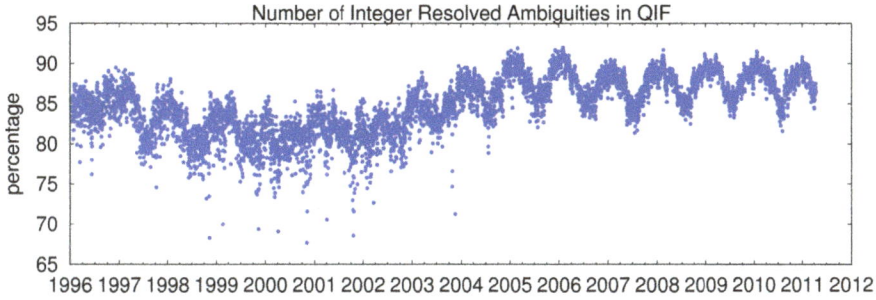

Figure 2: Success of integer ambiguity resolution over processed period

products before 2000. The results of daily and weekly solutions (normal equations) providing station coordinates were archived in the SINEX (Solution INdependent EXchange) format in order to provide official contributions to EPN-repro1 project and in Bernese binary (NQ0) format to support GOP long-term stacking of weekly normal equations.

2.2. Daily and weekly coordinate time-series

Daily and weekly solutions provide estimated coordinates of all processed stations in epochs of minimum variance (in ideal case equal to a central epoch of processed data period). Because the subnetwork has a regional extent only, a no-net translation (NNT) condition for a priori coordinates of selected fiducial stations was applied for datum definition. For each daily and weekly solutions, the individual set of consistent fiducial stations was iteratively selected from the a priori list of reference stations. This enabled us to visualize coordinate time-series in a single geodetic datum and to identify immediately most of the problems related to indidivual station performances. Figure 3 gives only a few examples for six stations: (a) GOPE – instrumentation change, (b) HFLK – seasonal variation probably due to a cumulating snow/ice on top of the antenna radome, (c) OBE2 – a poor data quality, (d) DRAG - station outside the Euroasian tectonic plate, (e) TUBI – non-linear post-seismic station movement and (f) MAR6 – post-glacial uplift in Fennoscandia. These plots were useful for the preparation of GOP long-term cumulative solution and in particular for the definition and revision of station coordinate and velocity discontinuities.

2.3. GOP long-term solution (1996-2011)

A 15-year solution was based on a cumulative combination of weekly normal equations, where only coordinates were kept in normal equations. The tropospheric parameters and ambiguities were pre-eliminated during weekly combinations. However, we cannot keep the coordinates constant over a long-term period and we need to apply at least a linear model (i.e. velocity) for each coordinate component. In some cases where non-linear trend in individual coordinate time-series exists (e.g. TUBI in post-seismic period, see Figure 3), a piece-wise linear model need to be applied.

The velocity parameters for all stations are thus introduced into the normal equations just before stacking weekly solutions and generating a unique cumulative one. The velocities can be estimated thanks to the applying long period during which the coordinates significantly

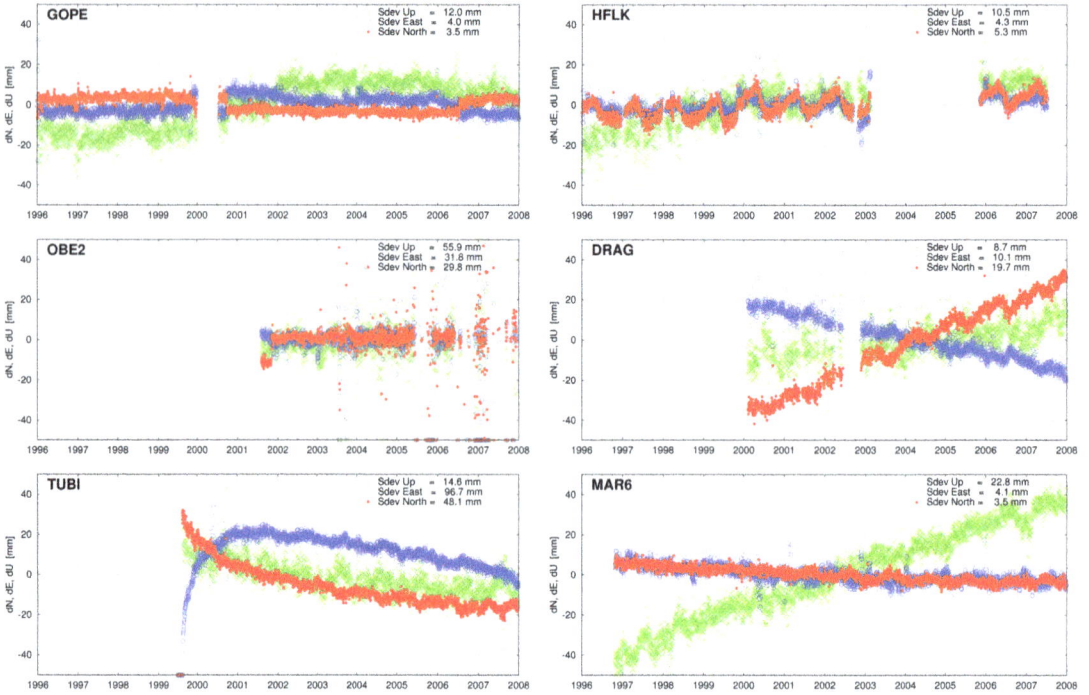

Figure 3: Example of coordinate time-series for six EPN stations from GOP daily solutions

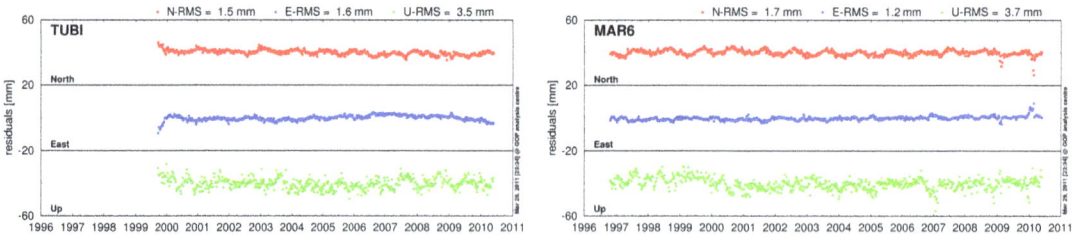

Figure 4: Cleaned coordinate time-series from the final cumulative solution

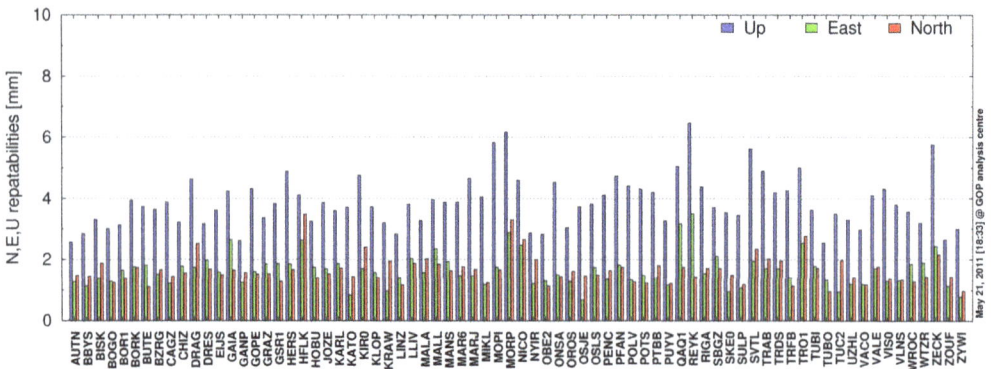

Figure 5: Coordinate repeatabilities of the GOP final cumulative solutions

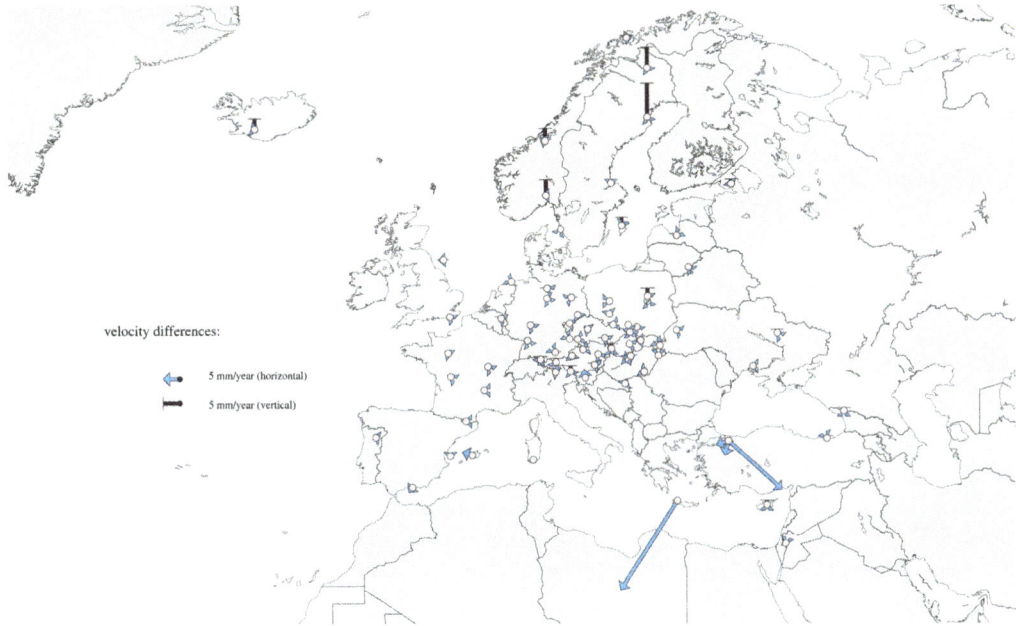

Figure 6: Residuals velocities with respect to ITRF2005

change (decorrelating the velocity estimation from the coordinates). For the stations with too short periods (usually less than a year), the velocities could be either constrained to some well-known a priori values or they are estimated with large formal errors showing correctly large uncertainties. A long-term solution was prepared based on the folowing steps, which might be individually re-iterated if necessary:

- 1st step : check residuals and indentify bad data periods
- 2nd step : setup discontinuities for coordinates and velocities
- 3rd step : select the best set of fiducial stations and provide final solution

A final solution is achieved after a careful cleaning of all poor data (usually identified as outliers or based on other external knowledges) and after setting discontinuities for all cases when station coordinates or velocities show significant jumps. Cleaned coordinate time-series (Figure 4) are results of such final solution, which already provides homogeneous repeatabilities for all station coordinates (Figure 5). In our solution, the repeatability was 1-3 mm and 4-6 mm for the horizontal and vertical component, respectively. The highest repeatabilities are clearly found at the station HFLK (on top of the Alp mountain) and those stations located at the margins of the network (stations QAQ1, REYK, MORP, TRO1, KIR0, SVTL, DRAG, NICO, TRAB, ZECK). This is due to their isolations (long-baselines) as well as less contributing observations. The datum of the GOP final solution was defined using a set of 18 stations. The maximum residuals after applying Helmert transformation (using translations only), between a priori and estimated coordinates, were below 5 mm in horizontal and 15 mm in a vertical components.

The estimated velocity differences with respect to the motion of the Eurasian tectonic plate (Figure 6) demonstrated stable changes of coordinates for most of the stations (usually below 1mm/year). This indicates a stability of the tectonic plate with respect to some others in

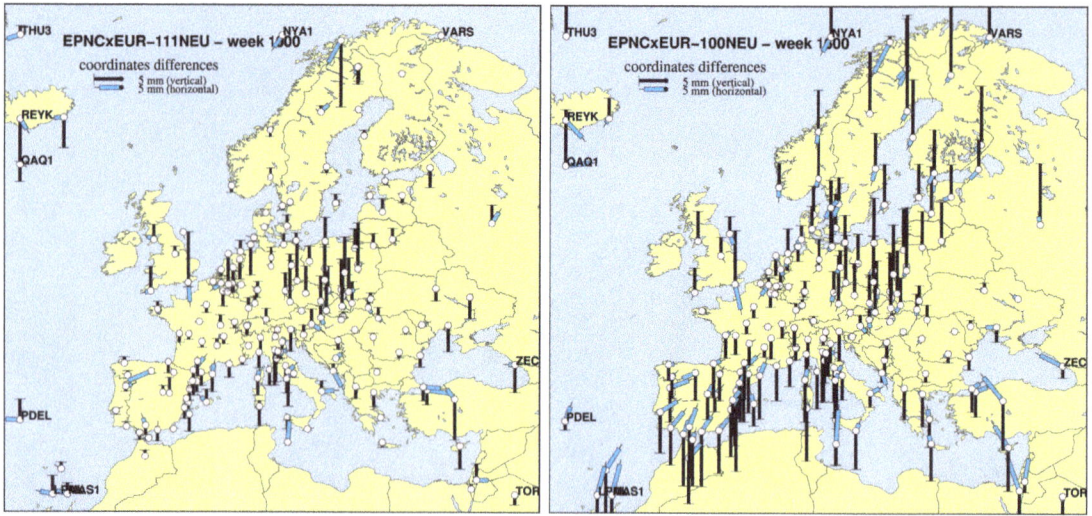

Figure 7: Residuals from the Helmert transformation at epoch of September, 7, 2010 (GPS week 1600) between the EPN cumulative solution (EPNC, updated at GPS week 1600) and EPN weekly solution (1600). Left: 7 parameters; Right: 3 translations (translation).

the world. However, we could clearly observe such geophysical phenomenas like: (a) post-glacial uplift in Fennoscandia, which is of different magnitude for various stations, (b) a single station in Greece representing a drift in the South–West direction common to most of the Greek islands, (c) non-linear post-seismic relaxation (modelled by piece-wise constant velocities) for TUBI station after the Izmit earthquake in 1999 (Mw 7.4).

3. Evaluation of European ITRF2005 densification

On November 5 2006, the IGS adopted new reference frame (ITRF2005) and new model for antenna phase center offsets and variations (PCV, IGS08) for its global orbit and clock products. This change caused jump at that epoch for almost all EPN station coordinates. The velocity estimation should not be teoretically affected if discontinuities are correctly setup and thus only velocities could provide a datum for the prediction of coordinates over a period of 3-6 years later (i.e. before a new ITRS realization was available).

The preparation and maintenance of the ETRS89 realization is one of the tasks of the EUREF Technical Working Group (TWG). Any further densification of ETRS89 on national level is also usually validated by the EUREF TWG. During 2010, several national densifications campaigns were validated: EUREF-2009-IR/UK, EUREF-2010-Czech and EUREF-2010-Serbia. Except EUREF-2010-Czech, which concerned a solution over three years [11], the others showed a North–South tilt from the coordinate residuals of Helmert transformation when applying (a) 3 translations + 3 rotations + scale or (b) 3 translations only. The reason was not known at that time, but the first impression was that the problem was not related to the campaigns themselves, but to the cumulative EPN solution providing a last realizion of the ETRS89. Consequently, an action item was raised to identify this problem.

We could use the GOP-repro1 as an existing historical homogeneous solution for more deep study, however, at the first stage we tested for which EPN stations the North–South tilt can be

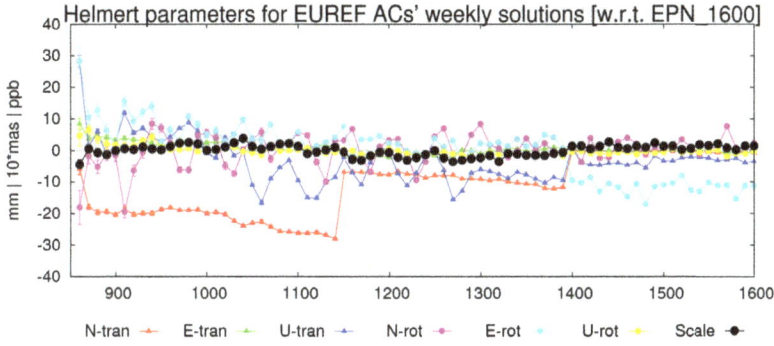

Figure 8: time-series of estimated Helmert parameters between EPN weekly and EPN cumulative solutions. All stations used; x-axis represents time in GPS week.

observed. We compared one of the latest combined weekly EPN solutions with respect to the latest update of the EPN cumulative solution. We used GPS week 1600 and all EPN active stations from which we iteratively selected a set of 'consistent' stations to be used as fiducials for the Helmert transformation. We applied 3 (translations) and 7 (translations, rotations and scale) parameters in the transformation between coordinates at a reference epoch in the middle of GPS week 1600 (September 8, 2010). Figure 7 shows the results, which clearly confirmed a common North–South tilt over all stations between the two solutions of about 2 mas (and additionally a small scale difference). This corresponds to about $\pm 1\,cm$ for heigh component for stations in North and South Europe for given epoch. Our expectation that the problem exists in the cumulutive solution was confirmed.

We have then run a similar test for every solution over a whole EPN period of GNSS observations (1996-2010) with a 10-week step. We should emphasize that the EPN cumulative solution provided coordinates, velocities and discontinuities at the common central epoch 2000.0, while the EPN weekly solutions provided coordinates at a central epoch of each week. Before applying Helmert transformations the coordinates of the EPN cumulative solution were firstly converted (by applying velocities) to the central epoch of EPN weekly solution. In order to provide a robust estimation of Helmert parameters, all available EPN stations were set a priori as fiducials while during an iterative procedure all those not consistent were temporarily excluded. Discontinuities defined in the EPN cumulative solution was necessary to take into account as well as to apply renaming scheme to compare theoretically identical stations only.

Time-series of 7 Helmert parameters between the two coordinate sets are depicted in Figure 8. The figure clearly shows jumps in the time-series due to existing inconsistencies in EUREF weekly solutions over the entire period. The inconsistencies are related to reference frame changes and different models applied (e.g. at GPS week 1400 antenna PCV model was switched from relative to absolute calibrations). The figure also shows time-series of Helmert parameters estimated in a local reference system centered at the geometrical centre of the EPN network. The North–South tilt, which is represented by the rotation around the East axis in the local system, has been clearly observed since GPS week 1400. In the global system the tilt is represented by X- and Z-translations and Y-rotation (not shown here). The figure was important to prove that we are able to reconstruct Helmert parameter time-series estimated

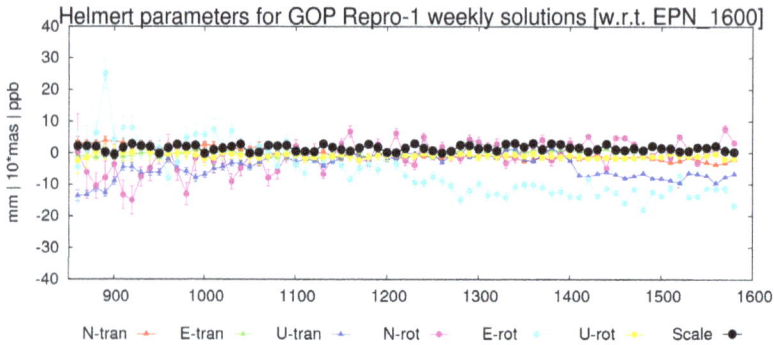

Figure 9: time-series of estimated Helmert parameters between GOP-repro1 and EPN cumulative solutions. GOP-repro1 stations used only; x-axis represents time in GPS week.

by the CATREF software [12] used during the combination of EPN weekly solutions into a cumulative one.

Additionally the figure demonstrated that our strategy developed for automated selection of fiducials was enough robust to achieve the same results as with datum uniquely defined during the EPN cumulative solution. The same strategy could be then used for the estimating tranformation parameters for GOP-repro1 weekly solutions with respect to the EPN cumulative solution and plotting corresponding time-series. Being a homogeneously reproccessing the GOP-repro1 solution enabled to reconstruct a full history of estimated Helmert parameters in a consistent way. Figure 9 shows such smooth time-series not affected by regular updates of reference frame or antenna calibration model.

The GOP-repro1 weekly products didn't include all EPN stations, but well covered the territory of Europe. We could guarantee the possible replacement of a complete EPN-repro1 solution, which was not available at that time, with the GOP-repro1 since reaching a good agreement of Helmert parameters estimated for both solutions after GPS week 1400. All comparisons before GPS weeks 1400 for GOP-repro1 (Figure 9) thus represents a successful reconstruction of the full historical development of Helmert parameters and also the development in the trend of the North–South tilt. From the clear linear trend, the tilt can be understood as a product of the weak velocity datum definition in original and European ITRF2005 densifications. And this further resulted in the North–South tilt increasing with the lenght of the coordinate prediction.

4. ITRF2008 discontinuities assessed using GOP cumulative solutions

The GOP-repro1 cumulative solution gave us an opportunity to assess the latest ITRS realization, ITRF2008. The motivation was raised again during the validation of national ETRS89 densification campaigns. The question was wheather the station REYK can be used as fiducial due to a significant coordinate jump in the ITRF2008. Such jump was, however, not visible in the EPN cumulative solution although an earthquake occured on September 11, 2008 close to the station.

The GOP-repro1 proved that the REYK position was not disrupted by the earthquake and relevant discontinuity in the coordinate time-series is not necessary. For visualization, we have developed a plot schemes showing clearly all instrumentation and reference frame changes at each station. Additionally, these plots include all setup discontinuities (coordinate offsets) as they were estimated in various reference frame realizations. Finally, the GOP-repro1 cumulative solution residuals are plotted to show all remaining (unmodeled) effects in the coordinate time-series. Figure 10 shows an example for station REYK with the specific jump (September 11, 2008) in coordinates from ITRF2008. The GOP-repro1 residuals before and after this epoch are smooth although a discontinuity was not setup for this epoch. Another station showing unrealistic discontinuities of several centimeters was TRDS (Figure 11) from those included in GOP-repro1 solution.

Besides discontinuities and their comparisons the plots clearly display the inconsistencies in station availability in various solutions. A limited validity of a coordinate prediction exists whenever the discontinuities were recently defined and observations from later period were not available. Figure 10 (REYK) and Figure 11 (TRDS) can represent typical, but not worst examples. Other examples are MAR6, RIGA and JOEN (not shown) providing an uninterrupted coordinate time-series over a whole period, but with missing first three years in ITRF2008.

The above assessments were important to demonstrate the necessity of the careful ITRF2008 station selection for their use as fiducials for regional or national densifications. It also shows some weak points of a global GNSS solution when considering aspect of regional permanent stations. The IGS-repro1, which was a basis for a global GNSS contribution to the ITRF2008, didn't provide the best solutions for at least the following cases: (1) for stations with missing data when compared to their official validity (and availability) within the EPN or, opposite, stations with data excluded from this period (both affecting the quality of estimated velocities) (2) for stations which discontinuties were not properly handled and, consequently, the station can not be considered as fiducial for this period or later (meaning that coordinates for specific periods were not correctly estimated).

5. Conclusion

The paper presented the GOP first reprocessing results, which officially contributed to the EPN-repro1 project. Additionally, the 15-year GOP cumulative solution was described providing long-term station coordinates, velocities and their discontinuity estimation. Repeatabilities estimated from cleaned long-term coordinate time-series reached 1-2 mm and 4-6 mm in horizontal and vertical component, respectively. We then showed the exploitation of GOP reprocessing results in the assessment of the EUREF ITRF2005 densification and the latest ITRS realization, ITRF2008.

We identified and confirmed the North-South tilt ($\approx 2mas$) in the currently available European reference frame based on the EPN ITRF2005 densification solution (e.g. last EPN cumulative solution updated in GPS week 1600. The study showed a historical development of the tilt and its close relation to a weak velocity datum definition in this frame. The velocity datum is particularly important for a long-term datum prediction when for most stations discontinuities were setup due to changes in processing strategy. The results of the tilt study were presented at the EUREF Technical Working Group meeting (Padua, March 2011) and confirmed by new European ITRF2008 densification already based on solutions from the EPN-repro1 project.

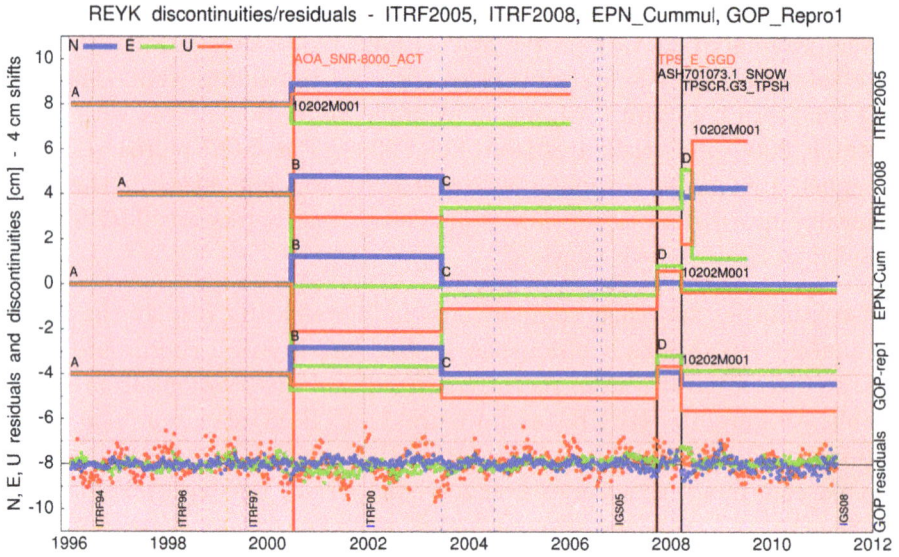

Figure 10: Coordinates discontinuities for station REYK for various solutions (ITRF2005, ITRF2008, EPN cumulative solution updated in 1600 and GOP-repro1 for which also residuals are shown). The changes in reference frames are marked in magenta besides the station specific instrumentation updates.

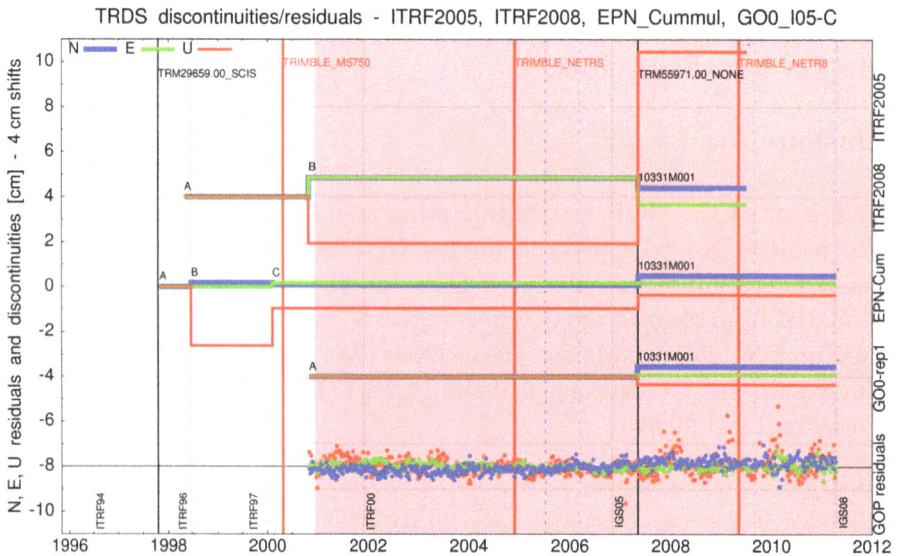

Figure 11: Coordinates discontinuities for station TRDS for various solutions (ITRF2005, ITRF2008, EPN cumulative solution updated in 1600 and GOP-repro1 for which also residuals are shown). The changes in reference frames are marked in magenta besides the station specific instrumentation updates.

Finally, selected EPN station coordinates, velocities and discontinuities of the latest ITRS realization (ITRF2008) were assessed using GOP-repro1 results. Specific problems for some EPN stations were identified in the global reference frame. This emphasized a further necessity of careful revision of all ITRF2008/IGS08 stations before their use for datum definition in regional densifications.

Acknowledgements

This work was partly supported by the Technology Agency of the Czech Republic (TB01CUZK006) and by the European Regional Development Fund (ERDF), project 'NTIS – New Technologies for Information Society', European Centre of Excellence, CZ.1.05/1.1.00/02.0090.

References

[1] Bruyninx C.: The EUREF Permanent Network: a multi-disciplinary network serving surveyors as well as scientists", GeoInformatics, Vol 7, pp. 32-35, 2004

[2] Gurtner W.: Guidelines for a Permanent EUREF GPS Network, In Report on the Symposium of the IAG Subcommission for the European Reference Frame (EUREF) held in Helsinki, 3-5 May 1995, Gubler, E. and H. Hornik (eds), Bayerische Akademie der Wissenschaften, Muenchen, EUREF publ. No 4, pp. 68–72, 1995

[3] IERS - International Earth Rotation and Reference Systems Service, http://www.iers.org

[4] Dow J., Neilan R.E., Rizos C.: The International GNSS Service in a changing landscape of Global Navigation Satellite Systems, J. Geod., Vol. 83, pp: 191-198, (2009), doi:10.1007/s00190-008-0300-3

[5] Steigenberger P., M. Rothacher, R. Dietrich, M. Fritsche, A. Ruelke, S. Vey (2006), Reprocessing of a global GPS network, J. Geophys. Res., Vol. 111, B05402 (2006), doi:10.1029/2005JB003747

[6] IGS-repro1 project: http://acc.igs.org/reprocess.html

[7] Douša, J., Ehrnsperger W.: Permanent GPS Data Processing at the BEK and the GOP EUREF Local Analysis Center, In Report on the Symposium of the IAG Subcommission for the European Reference Frame (EUREF) held in Sofia, 4-7 June 1997, Gubler, E. and H. Hornik (eds), Bayerische Akademie der Wissenschaften, Muenchen, EUREF publ. No 6, pp. 59–63, 1997

[8] Dach R., Beutler G,, Bock H., Fridez P., Gade A., Hugentobler U., Jaggi A., Meindl M., Mervart L., Prange L., Schaer S., Springer T., Urschl C., Walser P.: Bernese GPS Software Version 5.0, Astronomical Institute, University of Bern, Bern, Switzerland, 2007, User manual.

[9] IERS Conventions (2003): Dennis D. McCarthy and Gerard Petit (eds.), (IERS Technical Note 32) Frankfurt am Main: Verlag des Bundesamts fuer Kartographie und Geodaesie, 2004, 127 pp., ISBN 3-89888-884-3 (print version)

[10] Mervart L.:, Ambiguity resolution techniques in geodetic and geodynamic applications of the Global Positioning System. Geod. Geophys. Arb. Schweiz, No. 53, p. 155, 1995

[11] Douša J., Filler V., Kostelecký J.(jr), Kostelecký J., Šimek J: EUREF-Czech-2009 campaign, presented on the Symposium of the IAG Subcommission for Europe (EUREF) held in Gaevle, Sweden, 2–5 June, 2010

[12] Altamimi Z., Sillard P., Boucher C.: CATREF software: Combination and analysis of terrestrial reference frames. LAREG Technical Note SP08, Institut Geographique National, France, 2004.

## 11

Panoramic UAV Views for Landscape Heritage Analysis Integrated with Historical Maps Atlases

Raffaella Brumana, Daniela Oreni, Mario Alba, Luigi Barazzetti, Branka Cuca, and Marco Scaioni

Politecnico di Milano, piazza Leonardo Da Vinci 32, Milan, Italy
`raffaella.brumana@polimi.it`

Abstract

Analysis of landscape heritage and territorial transformations dedicated to its protection and preservation rely increasingly upon the contribution of integrated disciplines. In 2000 the European Landscape Convention established the necessity 'to integrate landscape into its regional and town planning policies and in its cultural, environmental, agricultural, social and economic policies'. Such articulated territorial dimension requires an approach able to consider multi-dimensional data and information from different spatial and temporal series, supporting territorial analysis and spatial planning under different points of view. Most of landscape representation instruments are based on 3D models based on top-down image/views, with still weak possibilities to reproduce views similar to the human eye or map surface development along preferential directions (e.g. water front views). A methodological approach of rediscovering the long tradition of historical water front view maps, itinerary maps and human eye maps perspective, could improve content decoding of cultural heritage with environmental dimension and its knowledge transfer to planners and citizens. The research here described experiments multiple view models which can simulate real scenarios at the height of observer or along view front. The paper investigates the possibilities of panoramic views simulation and reconstruction from images acquired by RC/UAV platforms and multisensory systems, testing orthoimage generation for landscape riparian areas and water front wiew representation, verifying the application of automatic algorithms for image orientation and DTM extraction (AtiPE, ATE) on such complex image models, identifying critical aspects for future development. The sample landscape portion along ancient water corridor, with stratified values of anthropogenic environment, shows the potentials of future achievement in supporting sustainable planning through technical water front view map and 3D panoramic views, for Environmental Impact Assessment (EIA) purposes and for the improvement of an acknowledged tourism within geo-atlas based on multi-dimensional and multi-temporal Spatial Data Infrastructures (SDI).

Keywords: landscape heritage, UAV, image orientation, panoramic views, historical maps

1. Introduction

The paper relates on-going experiments driven by the necessity to provide new scenarios for retrieving geospatial knowledge of territory and instruments capable of managing information to better addresses landscape heritage policies. In this sense, the advanced geo-web instruments can give a contribution to support the preservation of ancient traces of precious anthropogenic environment, partly recognizable within the context and to provide a better comprehension of the participated citizen culture of territory.

Experiences carried out within the Atl@nte geoportal (an on-line open-source atlas of historical cadasters and topographic maps of Lombardy, www.atlantestoricolombardia.it), have provided a valid ground to compare the current landscape with different historical layers. Furthermore, they ask to enhance the comprehension of such complex areas introducing innovative representation and rediscovering the semantic content potential of ancient views for anthropogenic landscape interpretation and identity recognizing process by the people, unavoidable elements during the preservation process. A significant and well representative case study in the Italian Alpine foothills (pre-Alpine region) is reported and discussed. The possibility of utilising image sequences gathered by Unmanned Aerial Vehicles (UAV) combined with existing Digital Terrain Models (DTM) was investigated in order to obtain 3D textured models for landscape analysis, especially in areas featuring strong vertical edges (built environment of hills, coast, mountains, and the like). A model helicopter, flying at few tens of metres from the ground, has been tested to acquire images all around a given point. The composition of the images to form panoramic views can be used to reproduce the 'human' point of view at a certain location. This kind of landscapes cannot be represented in a realistic way using airborne imagery which are normally applied for state-of-the-art landscape 3D modelling.

According to landscape policies in EU and Italian frameworks, the necessity to develop new tools for representation to be used for landscape planning simulation (Sect. 1) has been discussed. These include experimentation of RC/UAV images for panoramic views reconstruction (Sect. 2). A case study area has been selected to support ancient view simulation for scenario analysis and touristic valorisation of historic itinerary road (Sect. 3). The problem of image-orientation for non-conventional image acquisition dedicated to orthoimage projection of non-conventional views has been discussed relating to sensor system RC/UAV platform and GPS digital camera (Sect. 4). The promising potentials of this approach for the future strengthening of e-contents through web atlas and modern devices are illustrated in Section 5.

1.1. EU and Italian legislative framework for landscape planning and scenarios simulation

The Italian legislative framework on the protection of the landscape has been developed according to the European Landscape Convention [16], stipulated in Florence in 2000 and ratified by the Italian government in January 2006. Here the landscape of every country includes the historical, monumental and the natural characteristics of the territory, considered as part of the cultural heritage of all European citizens. Identity and recognition become two key features of the landscape quality and contribute to the formation and the increasing of the individual and social quality of life. The landscape thus becomes a resource for sustainable development of all countries, with the result that the whole territory must be considered in the plans and programs to enhance the landscape, with the attention directed not only to the 'exceptional' places, but also to the 'everyday life landscapes and degraded landscapes'.

Acting on the recommendations contained in the European Spatial Development Perspective

(ESDP), prepared by the EU in May 1999 in Potsdam, the EU Landscape Convention also imposes the need for each Country to integrate landscape into regional and town planning policies, in cultural, environmental, agricultural, social and economic policies, as well as in any other policies with possible direct or indirect impact on landscape. The objective is to formulate general application principles, strategies and guidelines that permit the taking of specific measures aimed at the protection, management and planning of landscapes. In this frame, the Code of cultural heritage and landscape (DL2 N.42/2004, [18]) stated the procedures for the landscape granting, assigning the tasks of monitoring projects to the Architectural Heritage and Landscape Office, the Italian Ministerial body in charge of protection, called upon to make an assessment on the project proposals. Its opinion has a binding force for the Regions, the agency delegated for the final authorisation, which in turn must ensure that the projects will respect the regulatory guidelines for provincial and municipal planning, as contained in the Regional Landscape Territorial Plan. The legislative objective, through this whole process of approval, is to lead transformation of landscape taking into account the morphology of the places, the scenic and environmental context, and the traces of their history, and not to overlap uncritically and brutally to existing landscape. The Code made compulsory the submission of the 'report on landscape' together with the application for authorization, both essential for the assessment, made by the competent authority, of the project in relation to the elements of landscape value, highlighting the impact of the projects on the landscape and the elements of mitigation and compensation required. Furthermore, landscapes along the water paths and coastal areas are considered as regions of a great strategic importance for the EU hence in this sense, INSPIRE directive [18] together with Shared Environment Information System (SEIS) are identified as main tools to facilitate the information flow in these areas.

Regarding the assessment of the landscape compatibility of the transformation proposed, it is prescribed to submit synthetic analyses of both state-of-the-art and the project. They must include not only descriptions, current or historical map extracts, but especially detailed simulations, made by 'realistic photo-modelling that includes an appropriate area around the project, calculated from the ratio of existing intervisibility'. The 'photographic representation' of the project area and of the landscape contest must be taken from accessible places and/or scenic routes, as specified by the Scottish Natural Heritage for landscape management Environmental Impact Assessment (EIA) [17]. It must include fronts, skylines and visual perspective from which the transformation is visible, with particular reference to high visibility areas (e.g. slopes, coasts). In this context, the development of innovative 3D metric representation of the landscape is an improvement to the traditional 2D photographic representations of panoramic views.

2. Sensors for Landscape Image Processing: State of Art

Development of RC/UAV sensors for documentation, inspection and surveying of Cultural Heritage is a field partially explored with success in Archeology [5], by inheriting the horizontal flight of aerial photogrammetry and image orientation. With reference to Landscape and Environmental Heritage domain, different problems have to be investigated in photogrammetric RC/UAV applications, regarding image acquisition and orientation for 3D reconstruction of landscape with a complex morphology, needing unconventional camera poses, vertical images or/and oblique images. The need of an accurate flight planning and control requires data acquisition to be carried out along a regular path. Indeed, photogrammetric surveys

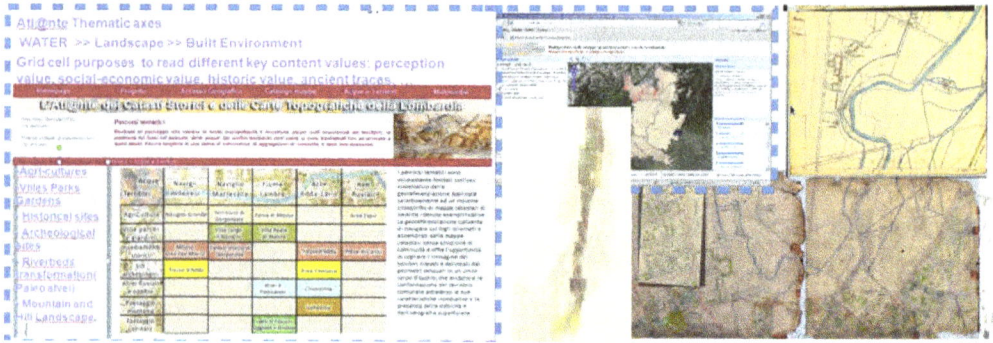

Figure 1: Atl@s web geoportal and the landscape area test along thematic axe of the Lambro river

need a block of images taken from different points of view at suitable scales and baselines. The integration with GNSS/INS [3,4,6] sensors allowed to obtain satisfying results, at the landscape scale. Camera calibration and image orientation have reached a high level in term of automation [8]. In the case nadiral images, the use uncontrolled flight path under the condition of well-targeted ground control points (GCP) enable the application of algorithms for automatic orientation. Research on fully automatic UAV image-based sensor orientation [8] opens opportunities in contexts like water view fronts devoted to metric image processing and information extraction.Moving the point of view, the GCP detection for the complex schema image block become difficult, especially in the case of very low cost uncontrolled flight, that ask to develop orientation and 3D modelling from markerless images [11] considering the wide spread potential of similar applications. Controlled sensor systems could be matched with the results obtained by multiple views 3D texturized models in the Computer Vision domain (Acute3D, http://acute3d.com/) in order to obtain technical instrument for line-of-sight analysis.

3. SDI Integration with Ancient View Simulation for Scenario Reconstruction and Touristic Valorisation of Historic Itineraries

Landscape heritage analysis, safeguarding and preserving can benefit from the integration of multi spatial-temporal data within a Spatial Data Infrastructure (SDI), rediscovering ancient map views of territory and different point of views of landscape observation [14]. The Atl@nte geoportal has provided a valid ground to deeper investigate and compare the current landscape with different stratified historical layers (Fig. 1) through the WMS/WFS generated on georeferenced ancient cadastral maps related to the current map (Google® imagery, 3D orthophoto maps, such as Terraitaly®, large scale technical map, etc.). Such a geo-metric grid allows to superimpose different chronological series, extracting information about transformations, permanences and mutations, especially if strictly related to the in situ data collection on the stratigraphic units, based on chrono-typological and archeometric approach [2].

A study case along water axes within Atl@nte was selected in the Italian pre-Alpine region. Motivation of this choice were the rich historic background and the complex landscape morphology of natural and built environment. Both aspects showed the potentials of providing new scenarios for retrieving geospatial knowledge of territory and instruments capable to support landscape heritage knowledge, safeguarding policies, addressing planning and divulgation instruments among citizens for sustainable tourism.

Rediscovering the perception of landscape value and synthesis represented in the past through observer point of views, generating '*vedute*' and perspective maps, water front views of portion of landscape faced on the inland water paths or lakes, that can be very useful if gradually integrated within modern geo-portal, introducing an added value of territorial map representation.

Architectural survey is usually obliged to move the point of view respect different plans of interest in function of the different structural elements (vertical fronts, facades, horizontal planes, floors, vaults). On the other hand, landscape representation is progressively trying to overcome the unique traditional nadiral privileged point of view represented by the aerial images acquired for current 2.5D cartography. Different point of views have to be gradually incorporated, as in the case of 3D city model navigation (e.g. Google Earth®) obtained by the individual projection of rectified images on building facades. The experimentation of mobile sensor platform image acquisition using an RC helicopter to reconstruct has been planned to reconstruct ancient panoramic views that represent an important witness of the great value assigned to this portion of landscape. The manifold aspects of interest will be addressed to in Section 1.2. In the conviction that rediscovery of ancient views can contribute to the rediscovery of the stratified values to preserve this area within planning policies and to valorise divulgation of touristic circuits, helping perception by the people using geo-portal, smart phone devices to strengthening comprehension of the historic geography of such places. In particular, a test was carried out on a simulation of ancient views generally acquired from privileged point of view (such as bell towers) on 'strategic' zones of territory, as in the case of 'non-rigorous' or quasi-perspective maps, such as the 'Pastoral Visits to the *Pievi*' by San Carlo Borromeo in Lombardy, XVI century (Fig. 3). Panoramic image acquisition by RC/UAV flight simulation has been experimented in order to verify the potentials and critical aspects for innovative representations of stratified multi-spatial/multi-temporal landscape reconstruction of complex strategic areas developed in the centuries along the ancient great itinerary roads. The sample test-site represents an important node along the Roman road from Aquileia to the North Europe, developed upon Neolithic settlement discovered around the area for favourable climate condition, water resources. This area featured a rich anthropogenic stratified territory with a natural defensive morphology. Celtic, Roman and Lombard archaeological settlement were found in the 'Park of Barro', and with the important Lombard-Benedictine Monastery of San Pietro al Monte, Civate. Despite the richness of this area we miss a synthetic view at 360^{circ} able to transfer those values to avoid the destroying intervention recently occurred by the new tunnel access exactly in correspondence of this node. An alternative solution, not completely far from the adopted one, would have more safeguarded this place.

A possible contribution to the divulgation of a participated citizen culture, by preserving ancient traces of precious landscape values conserved till now, can be given by advanced geo-web instruments.

Atl@s web can support historic geography and stratified ancient traces here represented on the traditional DTM and satellite imagery. The immediate perception given by the geospatial representation can be improved by privileged point of view of interest that can be usefully integrated to enhance the comprehension of beautiful natural and anthropogenic context

Figure 2: The sample landscape area, on the Lake of Annone closed to the access to the Lake of Como, is placed along an important node of the roman road from Aquileia to the North Europe. The principal stakeholder stratified landscape element are signalized.

Figure 3: Images acquired from ground level (left) do not provide an overall view of the ancient map (right). Panoramic images (centre) acquired at the height of the bell tower using RC platform flight, allowed to reconstruct the privileged point of view of XVI century Map of the Pastoral Visit to the Pieve by S. Carlo Borromeo acquired from the Bell Tower of the Church s. Eufemia and Romanic Baptistery.

unknown by the people at the local municipality and global level.

4. Non Controlled Flight and Controlled Sensor System for Landscape Orthoimages Generation from Panoramic Views

One of the aims of this project is the creation of panoramic images in order to texturize different landscapes. Since the approach is based on images taken from the ground instead of imagery used for aerial photogrammetric applications, some occlusions can be present. They can degrade the quality of the final results. This means that several objects located between the camera position and the investigated elements can be mapped generating a false result. On the other hand, this problem could be (partially) solved by employing images taken from different positions, with an interactive selection of the portions to be used.

Images acquired by an RC model helicopter have been used for high resolution modelling of the ground, tested on a portion of terrain with measured GCPs, while images acquired by digital camera integrated with GPS allows to solve orientation problem in order to obtain a 3D ortho-projection of a water view front portion without GCPs.

4.1. Flight test and Panoramic UAV images: data processing critical aspect

Automatic image orientation by AtiPE and automatic high resolution DTM extraction by using PMVS2 or LPS eATE techniques are here discussed with reference to the sensor system equipped on the platform and to the context. Furthermore, UAV images have been also used for high resolution modelling of the ground, tested on a portion of terrain with measured GCPs. In this experience, an automatic processing pipeline for the production of high resolution orthoimages has been setup. This includes automatic image orientation by AtiPE [8] and automatic high resolution DTM extraction by using PMVS2 or LPS eATE techniques. The former can be used to compute the exterior orientation (EO) through aerial triangulation of images taken with a calibrated camera, without any user interaction. In order to produce high resolution orthoimages or textured models from airborne sensors, terrain models derived from oriented UAV images using state-of-the-art matching techniques will be further investigated. Eventually, different software packages for orthoimage production from UAV data have been tested and evaluated according to image configuration and resolution. To check the metric accuracy achievable with the UAV system a photogrammetric block made up of 25 images was acquired in the area test over the municipality of Oggiono (Lombardy, Italy) on the top of a small hill with a dominant horizontal extension (Fig. 4). The UAV platform gives the opportunity to mount different types of cameras, according to the aim of each specific survey. In this case, a calibrated Nikon D90 (4288x2848 px) equipped with a 20 mm Sigma lens was employed. This camera has also the capability of acquiring video sequences. The network geometry includes images with a high overlap. Some of these were also quite oblique in order to obtain a better reconstruction of the elements over the horizontal plane (e.g. walls, columns, statues). The images were oriented with the ATiPE procedure [11], using SIFT features automatically extracted for several combinations of image pairs. To obtain a georeferenced result, 20 targets were placed on the ground. They were measured with a combined theodolite/GPS survey. 9 points were employed as GCPs in the bundle adjustment, while the remaining ones were used to check the quality of the estimated EO parameters. The coordinates of the centres of all targets were measured with the LSM algorithm, obtaining very precise images coordinates. The average height the above ground during the flight was ca 14m. RMSE values found were 5mm in the horizontal plane (x-y) and 17mm along the vertical direction (z).

Figure 5 shows a panoramic image generated by combining different pinhole shots. The object to be mapped is on the other side of the lake, facing the Roman road, the Benedictin Monastery and the Park of Barro. However, several other images were collected to obtain a final panorama with a large field of view, imaging also object in the nearby. A global visualization of the image can give nice results (although the weather conditions was not optimal due to mist). However, the final image is quite clean in correspondence of the target object, while for closer elements (e.g. trees, bell tower) some parallax errors were found.

4.2. Panoramic image for 3D Spatial Database texturing

The availability of an accurate 3D Spatial Database of the region allows to obtain a model usable for landscape representation purposes. This approach overcomes DTM extraction in order to generate panoramic photo projection of view front images acquired along riparian areas. Landscape visualization and photomontage experiments in the field of line-of-sight

Figure 4: The UAV system, an image of the area and the scheme of theodolite/GPS measurements.

Figure 5: The panoramic image generated from a model helicopter from the ancient view point.

map integration are beginning to implement visual impact assessment tools [12] in order to enhance metric approach with respect to qualitative approach and passive auto stitching software packages [1]. Advanced algorithms developed for textured 3D models on panoramic images acquired by controlled sensor have been utilised. Integration of GPS sensors with digital camera has simulated fully equipped flight control experimented in case of close range acquisition by ground instead of by flight; implementation of RC platform with this sensor system would allow to enhance the orientation of such geometry for orthoprojection purposes. Here is described a methodology for automatically texture a 3D Spatial Database by oriented panoramic image and the results of the test application carried out on the shore of Lake Como, at the appendix node of the area test. In order to obtain a more complete acquisition of the lake-front different images have been acquired for each standpoints (Fig. 6). Photogrammetric surveys was carried out with a full-format Nikon D700 (4256x2832 pixels) camera with calibrated 90mm lens. Each image was georeferenced with the so called 'photo-GPS' technique [9], where RTK-GPS measures of the camera positions and image are acquired together. For each position a set of images was acquired with 'photo-GPS' system, maintaining the same coordinates of image location while the orientation angles changed. Three panoramic images were generated from the images acquired at different stand points (Fig. 7).

The tie points of panoramic images were automatically extracted. Finally the camera positions and photogrammetric observations are adjusted together in order to obtain the EO parameters in WGS84 reference system. The oriented panoramic image has been automatically projected on the 3D Spatial Database (1:2,000 scale) and the first results are shown in Figure 8.

5. Conclusions

With reference to the sensor system, the implementation of controlled flight equipped on the platform with camera GPS allows to support panoramic image processing using automated

Figure 6: (Left) Test site, in red different standpoints and the lines-of-sight. (Right) 'Photo-GPS' system during the acquisition of images.

Figure 7: Two of the three Panoramic image acquired from different stand-points

techniques for image orientation and 3D surface reconstruction. Many critical aspects of the panoramic image projection of water coast views to obtain metric water front has to be faced and deeper investigated: segmentation of orthoimage projection on a 3D Spatial Database with the projection of portions of panoramic views along the sloped coast areas, analysis of geometric congruence and condition to synchronize the exact projection of the built up portion located on the second and third plane respect the foreground acquired on the close up building faced on the coast line. The research will continue the tests on these products, to support metric textured model generation and management within easy access software environment used by planners and professionals in order to generate a widely useful technical instrument. Developing multiple-choice representation would straighten the comprehension of complex scenarios, such as historic road and ecosystem corridors rediscovering new interpretation of the ancient itinerary maps and bird eye perspective contents. Modern devices integration to web geo-portal (such as smart phone, Ipad applications) could allow an on-site touristic divulgation and data sharing with local growth of sustainable touristic itineraries and identity process of knowledge and consciousness by the citizen. The sample area on landscape portion along water corridor with multiple stratified values of anthropogenic environment has shown the potentials of future achievements in this research field through a complementary integration of innovative technical map within atlas based on multi-temporal SDI: they contribute to enhance the comprehension of complex landscape portions, rediscovering the semantic content of ancient views for anthropic landscape analysis and identity recognizing process by the people, an unavoidable component in the preservation process.

References

[1] M.Brown, D.G. Lowe, 2005. Automatic Panoramic Image Stitching using Invariant Features, Int. Journal of Computer Vision, Volume 74, Number 1, 59-73, DOI: 10.1007/s11263-006-0002-3.

Figure 8: Figure. 8 (Above) Images of TopographicDB automatically textured by oriented panoramic images. (Below) VRLM model render images done within modelling sw environment (3DStudioMax)

[2] Brumana, R., Achille, C., Prandi, F., Oreni, D., (2006). 3D Data model for representing an historical Centre Site UDMS'06 25th Urban Data Management Symposium UDMS Aalborg, Denmark 1-8 9 Part IV

[3] Colomina, I. et al, 2007. The uVISION project for helicopter-UAV photogrammetry and remote-sensing. 7th Geomatic Week, Spain.

[4] Wang, J., 2008. Integration of GPS/INS/Vision Sensors to Navigate Unmanned Aerial

Vehicles. IAPRSSIS, 37(B1): 963-9.

[5] Bendea, H.F., Chiabrando, et altri, 2007. Mapping of archaeological areas using a low-cost UAV the Augusta Bagiennorum Test site. In Proc. XXI Int. CIPA Symp, Athens, Greece, on CDROM.

[6] Bento, M.D.F., 2008. Unmanned Aerial Vehicles: an Overview. InsideGNSS (Jan/Feb 2008): 54-61. [7] Eugster, H., S. Nebiker, 2008. UAV-Based Augmented Monitoring – Real-Time Georeferencing and Integration of Video Imagery with Virtual Globes. IAPRSSIS, 37(B1): 1229-1235

[7] Barazzetti L., Remondino F., Scaioni M., Brumana R., 2010. Fully Automatic UAV Image-Based Sensor Orientation. In Proc. of ISPRS Comm. I Symp., Calgary (Canada), IAPRSSIS 38(1), on CDROM, 6 pp.

[8] Forlani, G., Pinto, L., 2007. GPS-assisted adjustment of terrestrial blocks. In: Proc. of the 5th Int. Symp. on Mobile Mapping Technology (MMT'07). Padova, ISSN 1682-1777, CD-ROM, pp1-7.

[9] Wang, M., Bai, H., Hu, F., 2008. Automatic Texture Acquisition for 3D Model Using Oblique Aerial Images. First International Conference on Intelligent Networks and Intelligent Systems (ICINIS 2008), pp. 495-498, Wuhan, China.

[10] L. Barazzetti, F. Remondino, M. Scaioni (2010). Orientation and 3D modelling from markerless terrestrial images: combining accuracy with automation. The Photogrammetric Record, (pp. 356- 381).

[12] R. Berry, G. Higgs, M. Langford, R. Fry, 2010. An evaluation of online gas-based landscape and visual impact assessment tools and their potentials for enhancing public participation in the Uk, WebMGS 2010, XXXVIII-4/W13

[11] Remondino, F., Rizzi, A., 2010: Reality-based 3D documentation of natural and cultural heritage sites – Techniques, problems and examples. Applied Geomatics, Vol.2(3): 85-100

[12] Cuca B., Brumana R., Scaioni M., Oreni D. [2011], "Spatial Data Management of Temporal Map Series for Cultural and Environmental Heritage", International Journal of Spatial Data Infrastructure Research (IJSDIR) Vol. 6 (2011).

[13] M. Santana Quintero, K. Van Balen, (2009) "Rapid and cost-effective assessment for world heritage nominations", 22nd CIPA Symposium, October 11-15, 2009, Kyoto, Japan

[14] European Landscape Convention, Council of Europe, Florence, 20.X.2000. European Spatial Development Perspective (ESDP) - Postdam, May 1999. http://ec.europa.eu/regional_policy/sources/docoffic/official/reports/som_en.etm

[15] Scottish Natural Heritage (2008). http://www.snh.org.uk/publications/online/heritagemanagement/EIA/appendix1.shtml

[16] INSPIRE EU Directive (2007). Directive 2007/2/EC of the EU Parliament and of the Council (14 March 2007) establishing an Infrastructure for Spatial Information in the EU Community, Official Journal of the European Union, L 108/1(50), 25th April 2007.

## 12

Large geospatial images discovery: metadata model and technological framework

Lukáš Brůha

Department of Applied Geoinformatics and Cartography,
Charles University in Prague
Albertov 6, 128 43 Prague 2, Czech Republic
lukas.bruha@natur.cuni.cz

Abstract

The advancements in geospatial web technology triggered efforts for disclosure of valuable resources of historical collections. This paper focuses on the role of spatial data infrastructures (SDI) in such efforts. The work describes the interplay between SDI technologies and potential use cases in libraries such as cartographic heritage. The metadata model is introduced to link up the sources from these two distinct fields. To enhance the data search capabilities, the work focuses on the representation of the content-based metadata of raster images, which is the crucial prerequisite to target the search in a more effective way. The architecture of the prototype system for automatic raster data processing, storage, analysis and distribution is introduced. The architecture responds to the characteristics of input datasets, namely to the continuous flow of very large raster data and related metadata. Proposed solutions are illustrated by the case study of cartometric analysis of digitised early maps and related metadata encoding.

Keywords: Catalogue systems, SDI, image data, metadata, automation.

1. Introduction

From the perspective of spatial data infrastructures (SDI) development strategies, the old maps represent a specific data source, especially regarding metadata. Cartographic documents always represented a unique way of expression and distribution of spatial information. In addition to their aesthetic value, old maps also yield significant scientific value. In these historical documents the methods for the representation of space at the time of their origin can be revealed. Moreover, these records provide valuable historical spatial data and document the temporal changes of represented phenomena, and so information about land use changes, the development of towns or river network transformations can be discovered. Such pieces of information are in the center of interest of present-day geospatial information systems. The International Cartographic Association (ICA) has even established the Commission on Digital Technologies in Cartographic Heritage to promote the collections of early maps to a general public. The group focuses on the development of methodologies applied to archiving, accessibility and analysis of thematic and geometric content of old maps.

Traditionally, the libraries have been the institutions, which were responsible for custodianship and the availability of map products through organized collections in their conventional form (e.g. printed maps, journals, atlases, globes etc.). With the onset of digital era also the development strategies of libraries moved towards online distribution and constant request so as to provide improvable and innovative services for their users [27, 24]. Plenty of initiative has published early maps through geoportals, i.e. the David Rumsey Map Collection, the New York Public Library's instance of open-source Map Warper, the Geological Survey's Historical Topographic Map Explorer, the Office of Coast Survey's Historical Map & Chart Collection or the Alexandria Digital Library. Recently, some activities evolved aiming at the integration of digital map libraries with contemporary SDIs. Fernandez [10] presented an integrated access to the SDI through the crosswalk between geographic and bibliographic metadata profiles.

These issues regarding metadata interoperability and automation of data and metadata processing are also in the middle of interest of SDI solutions developers. The increase of spatial raster data volumes within SDI systems unveiled, how far the methods and tools originally designed for vector data are unsuitable for raster images, which are much bigger in data volumes and variable in storage formats [5, 22, 11]. The continuous flow of very large raster data from the digitization procedures of printed maps or satellite images receiving stations can be shown as an example of such data sources. These issues highly increase computational demands during any analytical task that requires additional data transfers, i.e. from the data store to the processing application. This implicates higher demands on the data store mechanism, cost-effective spatial images management and analysis approach. Also the availability of metadata, which supports support decisions on a dataset's appropriateness, alleviates the burden of users' requests on the system, allowing searching and assessing the data more effectively. The development of SDI is an active area of research regarding creation, update and authoring of such metadata [2, 26]. However, the development of the automation of metadata processing is still at the beginning and is being explored by many researchers [16, 18, 2, 14, 17, 21].

Accordingly, the solution presented here addresses the variability of metadata formats and differences between the standards used for data descriptions within libraries and geographical information systems (GIS). The model for metadata integration is introduced to link up the sources from these two distinct fields. This work describes the interplay between SDI technologies and potential use cases in digital libraries such as cartographic heritage.

Furthermore, this work proposes a technological solution focused on reduction of efforts related to the management of a continuous flow of large raster data and the creation of associated geospatial metadata. The architecture of the prototype system for automatic raster data processing, storage and distribution is introduced. This architecture responds to the characteristics of input datasets and to the demands for more effective spatial images management and analysis approach. Its characteristic feature is the shift of the application logic to the data store.

This article first discusses the metadata sources and proposes the metadata model for the integration of early maps metadata records within the SDI system. Further the technical and software requirements of SDI component for raster data are defined and the architecture, which is the basis for the system prototype implementation, is presented. Finally, the experiments and conclusions are reviewed.

2. Metadata model

Different standards with different focus are employed by librarian and SDI metadata catalogs. Currently, FGDC or ISO level standards are considered essential to spatial data infrastructures in order to satisfy the needs of archival, preservation and quality measure use cases. They are however, considered intricate and lengthy, without focus on discovery contexts such as web search or content description [12]. The insufficient metadata model for representation of the analysis outcome and the description of raster image properties can hinder reusability of existing analytical results and obstruct search for the desired image. In librarianship, the fundamental source of old map's description is represented by the bibliographic record. This description subsequently provides metadata for the digital environment usually encoded according to the MARC21 format. Such metadata is created in accordance with International Standard Bibliographic Description for Cartographic Materials (ISBD(CM)), Anglo-American Cataloging Rules with respect to local interpretation, i.e. AACR2R [1], or newer RDA [15] cataloging standard. Moreover, various cataloging procedures of old map collection [8, 19] and the final list of encoded metadata attributes are always depending on local specifics.

The survey of various metadata sources must be completed prior to the final constitution of the metadata model and the technological solution. Relevant information can be retrieved from multiple sources including raster headers, auxiliary files and external documents holding descriptive or technical metadata in a structured form.

2.1. Digital library and information science community metadata sources

Two kinds of metadata can be distinguished. First, the technical metadata describing the administrative and technical parameters of the digitization. Second, the actual bibliographic record entered mostly by the human description of the original document.

Technical metadata. The text documents of a well known format usually stored in an XML document are the source of technical metadata. These are usually produced in such a standardized format like Metadata for Images in XML schema (MIX) together with the image, during the extensive scanning campaign of archival materials. The owner of the original document, unique object identifier, data format, resolution, data volume and type of compression, identification and description of technical parameters of the scanner or the software used for scanning and post-processing belong among parameters encoded in such technical metadata documents. Metadata Encoding and Transmission Standard (METS) is a format frequently utilized by digital libraries to encapsulate logical connections between technical and bibliographical metadata.

Bibliographic records. The manual creation of some metadata like cataloguing procedure of early map collections cannot be avoided and represents an important source of metadata. Operators create metadata by writing descriptions of resources in a structured form, which can be automatically transformed afterwards. The bibliographical records are an essential resource for old maps description. During the cataloging procedure a plenty of descriptive attributes are recorded, including system identifiers, institution, authorship identification, title, physical description or scale, if available, and georeferencing entries etc. During the design of metadata model for discovery of geospatial resources, only a selection of all attributes were identified as relevant for such a purpose. The model focuses especially on elements describing the mathematical foundation of cartographic materials like the scale, projection or geographical extent, which are encoded in 034 and 255 fields of MARC21 format. An example of a map scale encoding in MARC21 including the conversion of obsolete measuring units [19]:

255 $a Měřítko [ca 1:49 000]. 15,9 cm = 3 wiener Zoll = 1 oster. Meile v. 4000 Klafter od. 10000 Schr.
255 $a Měřítko [ca 1 200 000]. 18,2 cm = 5 Meilen oder 20000 Klafter
255 $a Měřítko [ca 1 1 126 000]. 4,6 cm = 7 militaria germanica communia
034 $b 200000

For the sake of effective search for a desired digitized map, in addition to fields like author or period of creation, the content of the map and the time extent, which it is related to, the way of hypsography depiction or the language of the map are essential for cartographic document description. Through discussions with historical cartographers and geographers transpired the key role of proper representation of map series and map nomenclature, MARC21 fields 490 and 830. For instance The Third Military Survey of the Habsburg Empire, in which case the nomenclature is well established for searching the proper map sheets, moreover, it helps to avoid the language ambiguity between map sheets due to the change of map language during the survey. An example of the printed special map 1:75000 description [19]:

1101 $a Rakousko-Uhersko. $b Militargeographisches Institut $7 kn20060405003
24510 $a Josefstadt und Nachod. $p Zone 4 Col. XIV $h [kartografický dokument] / $c K. u. k. militar-geographisches Institut ; Geripp. Scherling ; Terrain Neumann
4901$a [Die Franzisco-Josephinische Landesaufnahme] 1 75 000
830 0 $a Třetí vojenské mapování 1 75 000 ; $v 4XIV, 1903.

2.2. Metadata model for early maps discovery

Proposed final crossing between metadata elements of the digital library sources and ISO 19115:2003, see Table 1, respects the methodology of digitization and cataloging [19] of an old map collection of Faculty of Sciences, Charles University in Prague, which provided the source of bibliographic metadata. Further, it respects the minimal requirements of ISO 19115:2003 standard and INSPIRE directive for a valid record and it is compliant with the standard for digitization of cartographic documents of National Digital Library of the Czech Republic [25].

The elements of the resulting metadata record of an early map can be divided into three classes, as is indicated in Table 1. First, the bibliographic records originally created by cataloging operators. The technical descriptions encapsulated in the METS document bearing information about the map's scan quality, resolution, compression and scanning conditions, as well as an object identifier to facilitate linking the records back to the digital library system are the second source. Among technical metadata elements, the parameters generated by the system itself, i.e. data source URL, are assigned, including some preset values common for the whole dataset of early maps (*Contact, Role* or *Metadata language*). Beyond these two groups of metadata fields, which originate from the automatic processing line of early maps scans and related metadata, the product of automated or manual map analysis can be encoded within the *Reference system* and *Reference scale* fields or within *Supplemental field* for other results.

Table 1: The model for creation of an early map metadata document.

Fields of ISO 19115:2003	Source of values - fields of MARC21 document
Title	830 av + 245 $a
Date	008 (character position 07-10, 11-14)
Abstract	If available: 520 $a Else: 245$b$c + 246$a + 300$a + 490$a + 500 $a
Keyword	043 $a + 600 $a + 610 $a
Language	008 (character position 35-37)
Topic category	650 $a + 651 $a
Temporal extent	648 1 $a (character position 1-4 a 5-9)
Geographic bounding box	034 $d, $e, $f, $g
	Source of values - technical metadata
Issue identification	METS document - field <ObjectIdentifier>
Identifier	UUID generated by metadata catalogue
Metadata language	"Cze"
URL	URL generated by map server
Metadata date stamp	System time of the document publishing
Organization name	Responsible party
Contact	Contact on responsible party
Role	Role of the responsible party
Hierarchy Level	"Dataset"
Statement (Lineage)	METS document fields <BasicImageInformation> + <ImageCaptureMetadata>
	Source of values - map analysis
Reference system	If available in MARC21 document: 255 $b Else: Cartometric analysis output - EPSG code
Reference scale	If available in MARC21 document: 034 $b Else: Cartometric analysis output - scale denominator
Supplemental information	Parameters of projection, scale and rotation If available: positional displacement, symbology type

3. Technical requirements for the implementation of SDI component for raster data

The automation of raster data processing, archiving, analysis and/or distribution requires the appropriate technological means. Following technical components were identified as essential for the design and implementation of the SDI module for raster image management:

- database system

- catalogue system

- visualization system

- administration system.

These components will be described in detail in following sections.

3.1. Database system

Within the database system the data model, which integrates both the data and metadata, is formed. Spatial database management system (SDBMS) capable of non-spatial data and georeferenced raster data storage is required to accommodate such a data model. Database tables would store raster datasets in a native format of the selected SDBMS platform, metadata tables would hold the ISO 19115:2003 elements. The current usual practice utilizes the out-of-the-database raster data storage, when only related metadata is stored in a relational database, while rasters are kept in the original raster binary format. The in-database strategy [20, 29] employed by the proposed solution stores images in a native database format, thus moving the image processing closer to the data and allowing for both concurrent, and parallel data processing. Database platforms with in-database raster data storage support also natively provide computationally optimized functions for raster data manipulation, editing and yielding of image statistics or image histogram. This functionality is re-usable within the data store and can be incorporated into the raster-based analytical procedures to be developed, making their implementation straightforward and their computational performance efficient. The functional requirements on SDBMS platform are summarized as follows:

- spatial indexing

- raster bands accessors

- raster pixel accessors and setters

- raster band statistics

- datum definition and coordinate system transformation

- storage of attribute data and metadata to the stored spatial data

- support for Geospatial Data Abstraction Library (GDAL) for raster format load and export operations

- raster pyramids and tiling support

- extrusion of raster regions

- storage of geo-referenced rasters and vector data

- analytical tool for detection of intersections with vector data

- map algebra over individual pixels.

The chosen database platform is also supposed to provide tools for communication with desktop and web applications. PostgreSQL with spatial extension PostGIS was chosen for the implementation of the proposed SDI concept. PostGIS raster extension supports raster data manipulation and complies with the stated requirements. This open-source solution is widely used and involves stable and active community. It ensures a future development and also easy cooperation with other academic institutions.

3.2. Catalogue system

Catalogue servers or metadata solutions, that include server side, are the second component required for the system design. The common objective for metadata catalogue usage within the SDI is to integrate all metadata records of disparate data sources and distribute them in a compact manner. The requirements laid on the catalogue are:

- the support of management and administration of the metadata

- the ready to use web graphic user interface (GUI) to interact with the end users

- extensibility of metadata profile templates

- the searching mechanism supporting the filtration based on multiple parameters

- metadata import via eXtensible Markup Language (XML) services.

GeoNetwork opensource catalogue is the commonly used catalogue solution. It fits to the requirements stated above and was finally selected for the integration with the prototype system proposed in this work.

3.3. Visualization system

The third key component of the proposed solution is the server technology capable of map distribution. This mapping application would be connected with the spatial database and would play the role of middleware by delivering the data to the client-side, providing on-line mapping service for the SDI. The requirements laid on the map server include:

- web administration environment for map layer management;

- Representational State Transfer Application Programming Interface (REST API) to programmatically manage the server;

- OGC compliant Web Map Service (WMS) support

- role-based access control to authorize users or groups of users.

Common use of the GeoNetwork metadata catalogue is in conjunction with the GeoServer map server, which act as a source of published georeferenced images (WMS) or complex features (Transaction Web Feature Service). Well documented and easy to deploy integrated system of GeoNetwork and GeoServer together with the fulfillment of the requirements stated above are behind the decision to employ the GeoServer as a map server application for the prototype solution.

3.4. Administration system

The design and implementation of an executable application would be another prerequisite to automation of the continuous flow of raster data and related metadata processing. Its objective would be to initialize and control the manipulation of raster data and provide an integrated approach to metadata generation. Operating system independence is the major requirement laid on such an application.

3.5. Architecture of the prototype solution

The architecture of the automatic raster data and metadata management system is depicted in Figure 1. This architecture is composed of three main layers:

- administration layer
- storage layer
- service layer.

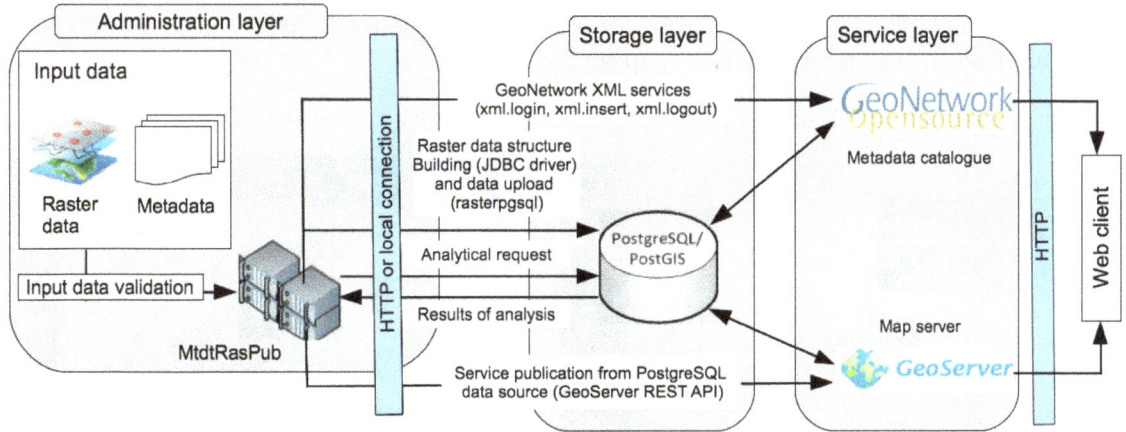

Figure 1: Architecture of the raster-based SDI module prototype system.

Administration layer

Administration layer provides an environment for initialization and configuration of the complete solution for automatic raster data and metadata processing and publishing. Technically, the administration layer is based on Java application (MtdtRasPub). The application continually checks for unprocessed data and triggers the building of database structures, the import of data and metadata to the database and the web map service publishing, when such data are detected. The administration layer is also responsible for management of related metadata. Procedures deployed within this layer take advantage of known structure of the metadata source documents and use xPath expressions to harvest the required elements defined in metadata model. The administering application executes requests for analysis of raster data. Such requests are sent to the data store, where the analytical logic is deployed. The in-database concept allows avoiding moving large data sets from the databases to detached analytical software. The results of analytical procedures or the metadata describing their output are received back.

Storage layer

The storage layer contains the databases, which store the spatial data and metadata. The PostgreSQL database platform is employed by the prototype solution.

Metadata database. According to the ISO 19115:2003 specification, the metadata records are stored in the XML structure. For the sake of automation of storage, update and publication of metadata together with the corresponding data, metadata needs to be stored in relational database. The metadata document is formed within the MtdtRasPub application and consequently GeoNetwork's XML services (xml.metadata.insert operation) are utilized for update or creation of new records in the catalogue.

Geospatial raster database. The raster data storage model utilizes the native PostGIS Raster format to support the data analysis provided by the database platform. The communication between the database and the map server is achieved by the use of GeoTools Image Mosaic JDBC plugin (IMJDBC), which requires the defining geometry for every raster tile placement. PostGIS Raster function ST_Envelope is applied to create the geometry. This way of storage allows for manipulation with all PostGIS raster functions and simultaneously the publication through GeoServer map server.

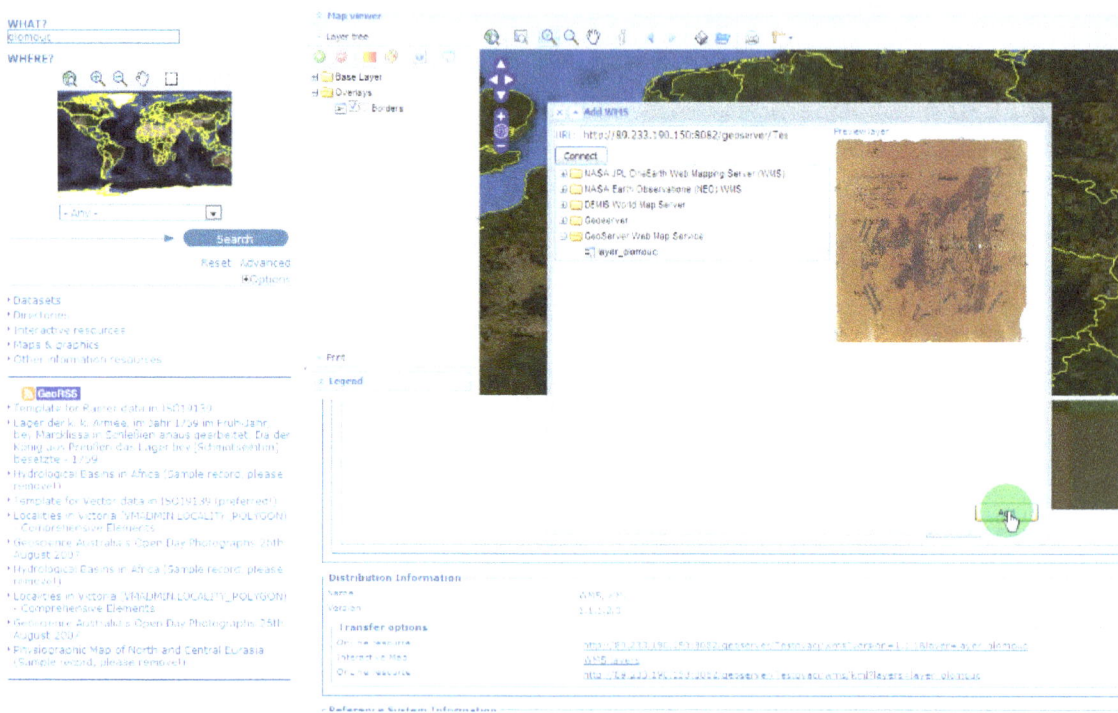

Figure 2: The overall view on the GeoNetwork's GUI with integrated metadata search, the overview of selected map's metadata and map visualization window.

Service layer

The objective of this service layer is to discover and publish metadata records and raster sources from PostgreSQL database through OGC compliant Web Services.

Metadata publishing. The deployed GeoNetwork opensource provides OGC Catalogue Services for the Web (CSW) server to access the metadata within PostgreSQL and enable search and update operations. This server supports 2.0.2 version of the OGC specification, supporting GetCapabilities, DescribeRecord, GetRecordById, GetRecords, Harvest and Transaction CSW operations.

Map service publishing. GeoServer supports as data source the rasters of three following data types - geotiff, worldimage and imagemosaic. To accommodate the needs of automatic publication of the data source retrieved from PostGIS Raster, the extension of supported data formats is necessary. The opensource nature of GeoServer allows adding among the supported formats the imagemosaicjdbc applying IMJDBC GeoTools plugin.

Graphic user interface

The architecture of the raster-based enhancement of SDI described above also provides a wide range of possible GUIs for user interaction with the prototype solution. GeoServer and GeoNetwork are modular open-source components following OGC standards and providing means for integration with desktop and web-based GIS systems, including web-based user interfaces. The GUI of GeoNetwork opensource, depicted in Figure 2, supports interactive user editing of ISO 19115:2003 metadata elements providing automatic updates to the database through CSW service. Integrated map window allows user to visualize the looked-up map and even combine it with another data available. GeoServer's GUI enables organization of individual map layers into group of layers for a signed-in user. These environments provide effective tools for constitution of map compositions from several layers as for the map purpose and user's needs.

Table 2: Metadata record of *Školní nástěnná mapa Afriky F. Umlaufta a J.G. Rothauga.*

Title	Školní nástěnná mapa Afriky F. Umlaufta a J.G. Rothauga
Date	1900
Abstract	pro české střední školy, upravil Josef Krejčí, 1 mapa, legenda
Keyword	regionální geografie; politická geografie; správní obvody; didaktika geografie; Afrika; nástěnné mapy; školní mapy
Language	Cze
Topic category	nástěnné mapy; školní mapy; Afrika
Temporal extent	1900
Geographic bounding box	-26.366944; 63.808333; -38.548056; 45.583056
Issue identification	1165964
Identifier	237
Metadata language	Cze
URL	http://mapy.natur.cuni.cz:8080/geonetwork/srv...
Metadata date stamp	2015-07-29T15:29:44
Organization name	CZ PrCU
Contact	mapcol@natur.cuni.cz
Role	point of contact
Hierarchy Level	Dataset
Statement (Lineage)	5250x4660; sRGB; IEC61966-2.1 2012-12-06; ScannTech 401i

4. Discussion and system evaluation

A solution for automatic management of continuous flow of large raster data and related metadata has been proposed. The presented technological framework allows to integrate descriptive information about images from various sources based on the metadata model and produces standardized geospatial metadata records. As this work focused on early maps, the metadata model to represent the corresponding map's description was introduced based on recommendations from librarians and cataloging operators. Table 2 shows a metadata record in ISO 19115:2003 standard of one particular dataset.

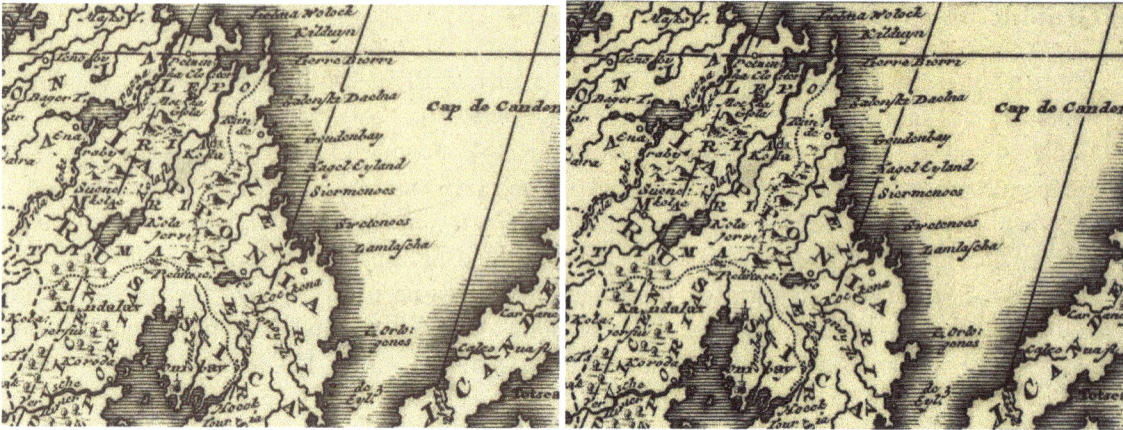

Figure 3: The segment of an early map (a) before and (b) after application of image histogram stretching procedure.

The technological solution of the prototype system is based solely on opensource technologies. While this approach has several advantages compared to proprietary systems, including the opportunities for customization and a support of active communities of developers, it also causes a number of technological challenges. Due to the heterogeneity of independently developed components, which were not implemented with mutual respect, the customization of the system components or additional libraries are required to enable the communication between all layers. That complicates the final configuration of the server solution. Another potential complication, identified also in works concerned with representation of metadata related to vector data [21], is the GeoNetwork's storage mechanism of metadata. The whole XML string is stored in a single column, thus accessing a single metadata element can only be done by GeoNetwork's parsing mechanism.

The consideration, whether the out-of-the-database or in-database raster data storage should be employed, must consider the major objective of such application. If the solution is meant for mere distribution of raster data, the out-of-the-database storage provides slightly better time responses and poses less complications when building the server solution and communication channel between GeoServer and PostgreSQL/PostGIS. Because the leading motivation was such a system, that would support the raster images analysis providing native database functions for raster data manipulation, editing and image statistics, the in-database approach to the data storage was chosen. This facilitated the development of new analytical procedures within the data store.

Image enhancement. Some tests of the in-database approach, like an image histogram equalization or cloud and snow detection from satellite imagery procedures, were presented in previous author's work [5]. The histogram equalization can be applied to improve the legibility of an old map or to remove differences in contrast between separate map sheets of map series for the sake of mosaicking, see Figure 3 for an example. Consequently, in the metadata field *Supplemental information* new intensity values of the map sheet can be retained.

As the approach focuses on digitized early maps and related metadata, the character of early map (fragility of the original and its uniqueness regarding the means of map representation or map construction) however suggests the human mediation in the creation of descriptions of some raster sources cannot be completely avoided. Nevertheless, the proposed metadata model facilitated to represent analytical results obtained manually or with a use of an external application.

Cartometric analysis. The cartometric analysis of an old map is another use-case, whose results are desired to be recorded for the future discovery. Cartometric analysis is a key pre-requisite for any proper analysis of a map content. Many authors, [7, 6, 28] to name a few, paid attention to development of cartometric analysis methods, so as to allow for environmental research, land use changes description and many other discoveries based on old maps. This experiment was focused on the assessment of planimetric accuracy and the estimation of cartographic parameters of old maps. Detected projection, category of projection, geographical coordinates of the cartographic pole, the latitude of the true parallel, the longitude of the prime meridian or percentage of points fitting used geometrical model are among parameters examined by the procedure. Theoretical background to the applied methodology is available in publications [23, 4]. The cartometric analysis procedure was performed by external software package detectproj [3]. Other parameters of map analysis, i.e. map symbology description, completeness of map content or positional accuracy can be done manually [13]. The deviation of the graticule of Delisle map [9] and the graticule of estimated stereographic projection is depicted by Figure 4.

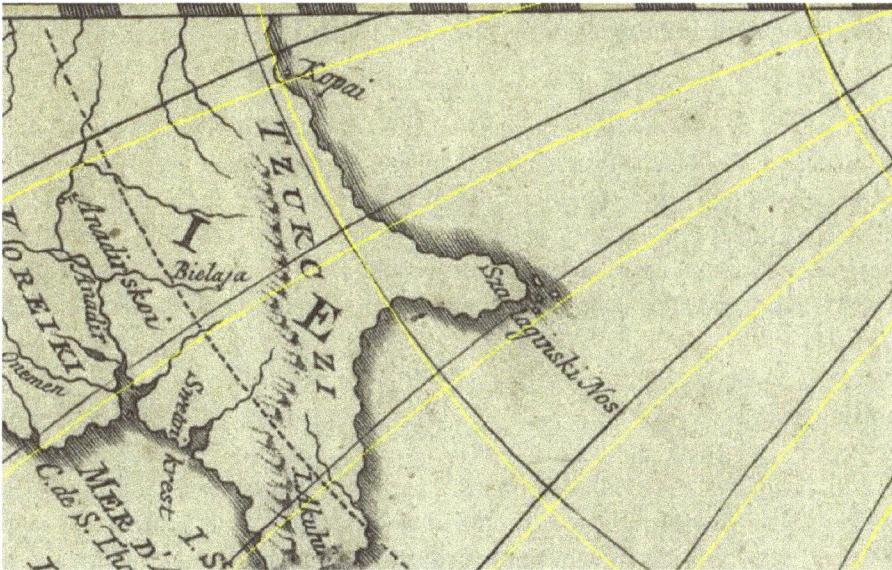

Figure 4: The visualization of the deviation between the graticule of Delisle (1774) map and the graticule of estimated stereographic projection. Figure drawn by [13].

Another example illustrates the measurement of positional displacements of selected cities, which is depicted in Figure 5.

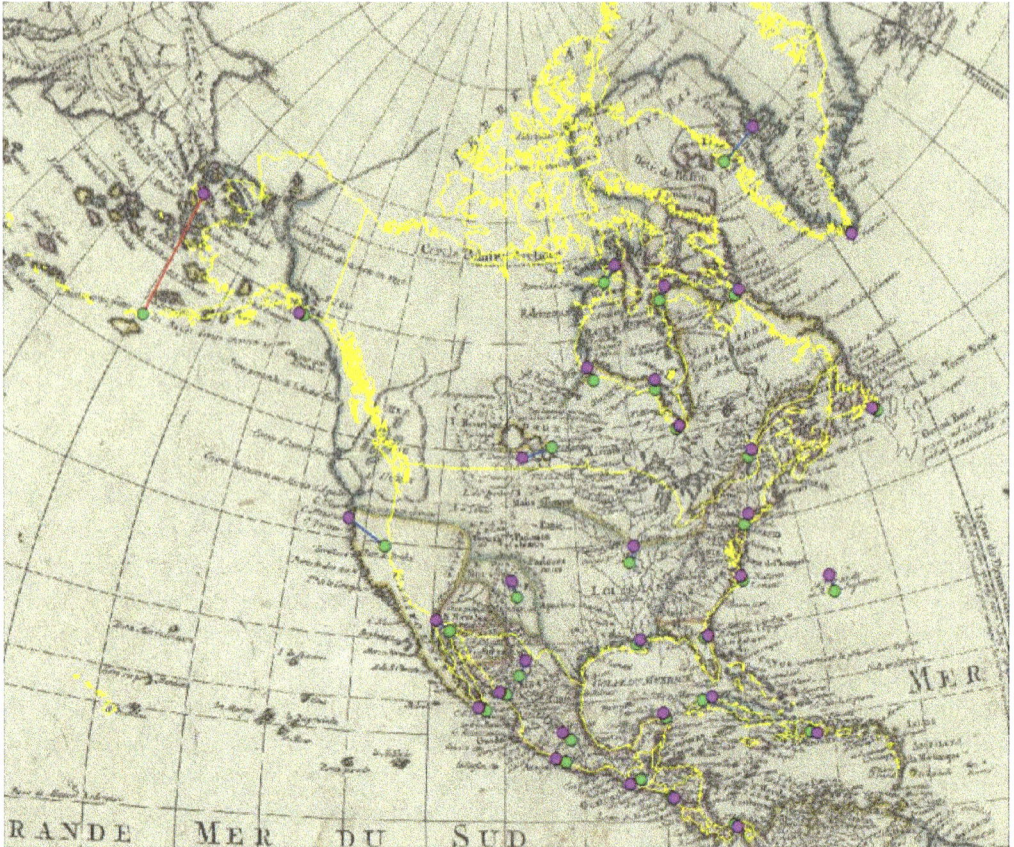

Figure 5: The illustration of the shift of points and the continent outline in Delisle map in comparison to the actual state. Figure drawn by [13].

The content of the *Supplemental information* field related to the Delisle map analysis and its visualization by the GeoNetwork metadata catalogue is demonstrated in Figure 6. The most probable detected projection, latitude of the true parallel, the longitude of the prime meridian, geographical coordinates of the cartographic pole, HOMT - standard deviation on identical points after homothetic transformation, HELT - standard deviation on identical points after Helmert transformation and measured positional displacements on selected places were encoded.

5. Conclusion

Efforts for disclosure of map collections are one of the very actual aspects of the SDI use. Moreover, it is also an impulse for further development of metadata interoperability and tools for effective processing of continuous data-flows of raster datasets generally. These efforts were presented through the example of the map collection of the Charles University in Prague and the SDI of the Faculty of Science. The interplay between SDI technologies and librarianship was described in this work and presented on some examples of a cartographic heritage discovery.

Introduced solution met the needs for automation of a continuous flow of very large raster data

```
Projection & Scale analysis:
1 stere 0.0 -89.5 0.0 0.0:
Scale HOMT: 42865490.7
Scale HELT: 42865587.5
Rotation HELT: -0.12 deg
----------------------------
Positional displacement [km]:
Ciudad de Panamá 48,91; Cape Hatteras 58,29; León 26,76;
Bermuda 186,23; Ciudad de Guatemala 153,30;
New York City 119,18; Acapulco 130,65; Québec 85,50;
Ciudad de México 145,64; Cape Race 45,80; Cabo Catoche 53,48;
Dauphin 357,84; La Habana 117,50; York Factory150,40;
Santo Domingo 76,14; St. Augustine 45,90; New Orleans 54,29;
Unalaska Island 1130,73; Santa Fe 202,70; Cape Mendocino 532,35
```

Figure 6: The content of *Supplemental information* field within GeoNetwork catalogue.

and associated metadata. And, as a result, it saved custodians' time and limited the space for manually introduced errors. The released human resources can concentrate on descriptive metadata creation to support semantic search and enhance the raster data discovery. The actual efforts aim at server-based deployment of the solution. The future development founded on proposed technology and architecture will be focused at the extension of content-based image analyses followed by automated content-based metadata creation.

Acknowledgement

This work has been supported by the TEMAP project of the NAKI program of the Ministry of Culture of the Czech Republic with DF11P01OVV003 code.

References

[1] Ivana Andresová. *Anglo-americká katalogizační pravidla: příručka pro katalogizátora s příklady ve formátu UNIMARC a MARC 21*. Praha: Národní knihovna české republiky, 1994. ISBN: 80-705-0187-1.

[2] James K. Batcheller. "Automating geospatial metadata generation—An integrated data management and documentation approach". In: *Computers & Geosciences* 34.4 (Apr. 2008), pp. 387–398. DOI: 10.1016/j.cageo.2007.04.001.

[3] Tomáš Bayer. *Detectproj: estimation of the cartographic projection and its paramameters from a map*. Software. 2015.

[4] Tomáš Bayer. "Estimation of an unknown cartographic projection and its parameters from the map". In: *GeoInformatica* 18.3 (Mar. 2014), pp. 621–669. DOI: 10.1007/s10707-013-0200-4.

[5] Lukáš Brůha. "Automation of geospatial raster data analysis and metadata updating: an in-database approach". In: *GEOGRAPHICA* 49.2 (Nov. 2014), pp. 49–56. DOI: 10.14712/23361980.2014.14.

[6]   Jiří Cajthaml and Jiří Krejčí. "Využití starých map pro výzkum krajiny". In: *Sborník sympozia GIS Ostrava 2008*. Ostrava: VŠB-TU Ostrava, 2008.

[7]   Jiří Cajthaml et al. "Georeferencing and Cartographic Analysis of Historical Military Mappings of Bohemia, Moravia and Silesia". In: *CTU Reports, Proceedings of Workshop 2007*. Praha: Czech Technical University, 2007.

[8]   Joan Capdevila et al. "Gateway MARC21-ISO19115: definition and reference implementation". In: *e-Perimetron* 7.3 (2012), pp. 155–162.

[9]   Guillaume Delisle. *Carte d'Amerique: divisees en ses principales parties*. Amsterdam: Covens & Mortier & Covens junior. 1774.

[10]   Alberto Fernández-Wyttenbach et al. "First approaches to the usability of digital map libraries." In: *e-Perimetron* 3.2 (2008), pp. 63–76.

[11]   Stanislav Grill et al. "Archive and catalogue system for receiving satellite data as a part of academic SDI". In: *Imagin[e,g] Europe*. Ed. by Manakos Ioannis and Kalaitzidis Chariton. 2010.

[12]   Darren Hardy and Kim Durante. "A Metadata Schema for Geospatial Resource Discovery Use Cases". In: *Code4Lib* 25 (2014).

[13]   Petra Jílková. "Severní Amerika ve starých mapách". Bachelor Thesis. Charles University in Prague, Czech Republic, 2014.

[14]   Mohsen Kalantari, Abbas Rajabifard, and Hamed Olfat. "Spatial metadata automation: a new approach". In: *Spatial science conference*. Ed. by et al. B. Ostendorf. Adelaide, Australia: Surveying & Spatial Sciences Institute, 2009, pp. 629–635.

[15]   Judith A. Kuhagen. *RDA essentials*. 2010.

[16]   Miguel Manso et al. "Automatic metadata extraction from geographic information". In: *AGILE 2004: 7th conference on geographic information science: conference proceedings*. Ed. by Toppen F. Heraklion: Crete Univ. Press, 2004.

[17]   Miguel Manso-Callejo, Mónica Wachowicz, and Miguel Bernabé-Poveda. "The Design of an Automated Workflow for Metadata Generation". In: *Metadata and Semantic Research*. Springer Science + Business Media, 2010, pp. 275–287. DOI: 10.1007/978-3-642-16552-8_25.

[18]   J. Nogueras-Iso et al. "Metadata standard interoperability: application in the geographic information domain". In: *Computers, Environment and Urban Systems* 28.6 (Nov. 2004), pp. 611–634. DOI: 10.1016/j.compenvurbsys.2003.12.004.

[19]   Eva Novotná. "Staré mapy a grafiky v Geografické bibliografii ČR on-line". In: *Knihovna : knihovnická revue* 24.1 (2013), pp. 5–27.

[20]   O. R. Obe and L. Hsu. *PostGIS 2.0 3D and Raster support enhancements*. (Visited on 08/17/2015).

[21]   Hamed Olfat et al. "A GML-based approach to automate spatial metadata updating". In: *International Journal of Geographical Information Science* 27.2 (Feb. 2013), pp. 231–250. DOI: 10.1080/13658816.2012.678853.

[22] Du A. Peijun and Chen B. Yunhao. *ISPRS Workshop on Service and Application of Spatial Data Infrastructure, XXXVI(4/W6), Oct.14-16, Hangzhou, China. Some key techniques on updating spatial data infrastructure by satellite remote sensing imagery.* 2005.

[23] Markéta Potůčková and Tomáš Bayer. "Application of e-learning in the TEMAP project". In: *Geoinformatics FCE CTU* 9 (Dec. 2012), pp. 91–99. DOI: 10.14311/gi.9.8.

[24] Muthu Sabarish Senthilnathan, Gkadolou Eleni, and Stefanakis Emmanuel. "Historical Map Collections on Geospatial Web". In: *GEOMATICA* 67.3 (2013), pp. 163–171. DOI: 10.5623/cig2013-035.

[25] Pavla Švástová and Jaroslav Kvasnica. *Definice metadatových formátů pro digitalizaci monografických dokumentů (monografií, kartografických dokumentů, hudebnin).* Tech. rep. Praha: Národní knihovna, 2013.

[26] Sergi Trilles et al. "Assisted generation and publication of geospatial metadata". In: *Proc.15th AGILE Int. Conf. Geographic Information Science: bridging the geographic information sciences.* New York: Springer, 2012, pp. 24–27.

[27] Ifigenia Vardakosta and Sarantos Kapidakis. "The New Trends for Librarians in Management of Geographic Information". In: *Procedia - Social and Behavioral Sciences* 73 (Feb. 2013), pp. 794–801. DOI: 10.1016/j.sbspro.2013.02.120.

[28] Bohuslav Veverka and Lucie Šrajerová. "Kartometrická analýza polohopisné přesnosti geografického obsahu historické Komenského mapy Moravy". In: *Sborník 18. kartografické konference.* Olomouc, 2008.

[29] Qingyun Xie et al. "In-database image processing in Oracle Spatial GeoRaster". In: *ASPRS 2013 Annual Conference.* Baltimore, Maryland., 2013.

## 13

Introducing the new GRASS module g.infer for data-driven rule-based applications

Peter Löwe

Helmholtz Centre Potsdam
GFZ German Research Centre for Geosciences

ploewe@gfz-potsdam.de

Abstract

This paper introduces the new GRASS GIS add-on module g.infer. The module enables rule-based analysis and workflow management in GRASS GIS, via data-driven inference processes based on the expert system shell CLIPS. The paper discusses the theoretical and developmental background that will help prepare the reader to use the module for Knowledge Engineering applications. In addition, potential application scenarios are sketched out, ranging from the rule-driven formulation of nontrivial GIS-classification tasks and GIS workflows to ontology management and intelligent software agents.

Keywords: module g.infer, GRASS, data-driven rule-based application

1. Introduction

Maps are used to represent the world surrounding us. They are put into use as tools to categorize, classify and judge our environments, to make decisions and act accordingly. In more general terms, the science of mapmaking, cartography, is to provide usable and understandable spatial information for a section of space for decision support. The motivation for computer driven cartography, mostly shouldered by Geographic Information Systems (GIS) such as GRASS GIS, is to perform the overall tasks of mapmaking as a workflow, including means to apply the human expertise and know-how required to infer decision-support for human actions. In GIS, the development of such „map-making" workflows is usually handled by stepwise execution of the consecutive processing steps by a human operator, to create and document the unfolding workflow, by interacting with the actual spatial data. Once a mapping workflow has been laid out, the next step is its automatisation, turning it into software. This can involve scripting, i.e. the definition of an execution-chain of available GIS modules, or programming, which includes the development of new GIS modules. Free and Open Source GIS like GRASS GIS allow rapid development of both solutions as the overall codebase can be exploited, so there is no need to reinvent previously developed functionalities because of copyright infringement issues. However, if a mapping workflow can be formulated by the human GIS operator, but can not be implemented as script or GIS module, there's a problem. In this case, the task at hand is basically solveable, but the available software environment lacks the flexibility to accommodate the workflow within acceptable time and effort constraints. This situation occurs frequently for classification tasks (remote sensing data or similar fields), resulting in the use of suboptimal classification algorithms: The implemented solution is not oriented on the original task solving strategy, but is limited by available software tools and programming skills.

A similar field is the flexible set up of GIS workflows which needs to adapt the processing chain according to changing constraints such as the availability and quality of data input, again within acceptable time and effort constraints. What is needed in these two scenarios, both for classification and GIS-workflow execution, is a way to encode understood, yet hard to formulate, „rules of thumb", acquired from human domain experts. A tool based on such rules will excel in scenarios where the following tell-tale indicators exist [28]

- Classification tasks which may not appear demanding, but no robust way for building a solution can be defined in acceptable time and effort.

- Simple workflows, where the processing rules keep changing depending on the available input and other parameters.

- Problems which have not fully understood or are very complex to solve.

In such cases, a rule-based approach, as provided by the new GRASS module g.infer becomes advantageous:

- Rule-based modelling allows to focus on "What to do" instead of "How to do it".

- Rules allow to express solutions to complex problems and to verify them consequently by logging the decision steps leading to a particular solution.

- Separation of logic (know-how) and data allows to keep the know-how to be stored in centralized rule-bases, providing a central point of access for further editing and improvement.

- Rules can also be human-readable, doubling as their own documentation and to be reviewable by domain experts.

While GRASS GIS provides a wide range of classification tools for raster, vector and volume data, a flexible yet generic approach tailored to conveniently express such rules, applicable to all GIS data types, is currently lacking. GRASS 4.x and GRASS 5.x featured the r.infer module, which provided basic rule-based analysis capabilities for raster data [20][23]. The module remains to be ported into GRASS6 and GRASS7. The same holds true for the r.binfer module , which uses an inference engine based on Bayesian statistics (making decisions based on past experience) to assist human experts in a field develop computerized expert systems for land use planning and management, basing bases the probable impacts of a future land use action on the conditional probabilities about the impact of similar past actions [1][2].

2. Artificial Intelligence

Artificial Intelligence (A.I.) is the field of computer science which focuses on the processes of human thinking an their implementation in software. A.I. is divided in various disciplines such as Artificial Life, Software Agents, Neural Networks, Genetic Algorithms, Decision Trees, Frame Systems and Expert Systems [6][25]. Several GRASS GIS modules, including the add-on modules r.fuzzy.* [13][14], r.agent[22], magical [16][17][18], ann.*[24], have been developed to solve tasks related to A.I. disciplines.

Knowledge representation and Classification is the A.I. Discipline focusing on how human knowledge for problem solving can be represented, manipulated and preserved. So called Knowledge-based Systems, also known as Expert Systems, facilitate the encoding of knowledge for automated reasoning or inference, i.e., the processing of data to infer conclusions, which can be mapped out in a GIS. The overall process of making human expertise available through an Expert System is called Knowledge Engineering. A Rule Engine implementing an Expert System instance for a specific knowledge-domain is called a Production Rule System. The term "Production Rule" stems from formal grammars where it is described as an abstract structure that describes a formal language precisely [15]. Such a Production Rule System is the core of g.infer.

3. The C Language Integrated Production System

The C Language Integrated Production System (CLIPS) is a Production Rule System toolkit for building Expert Systems. The project was started by NASA (Johnson Space Center) in 1985, where it was maintained until the 1990s. It is currently hosted at Sourceforge and is provided under a public domain licence. The name is an acronym for "C Language Integrated Production System", succeeding the original name „NASA's AI Language (NAIL)"[26].

CLIPS is written in C and provides a rule-based data-driven programming language. It resembles in syntax and user interface closely the language LISP [10]. CLIPS traces its origins to Inference's ART which in turn stems from OPS5 [27]. The CLIPS language continued to evolve and includes today paradigms for rule-based, procedural, functional and object-oriented programming. Since 1991, CLIPS includes the CLIPS Object Oriented Language (COOL) for object-oriented development. In the recent CLIPS Version 6.3, rules can be triggered via Objects enabling Object Oriented Modelling to drive the inference process.

Rules in CLIPS can be loose- or close-coupled: In the latter case, the activation („firing") of a rule explicitly invokes the firing of other rules. This is also referred to as a categoric problem solving approach [26]. On the other hand, loose coupling is achieved by a rule-base manipulating sets of variables or non-ordered facts, with independent rules pattern-matching on these, firing only if certain value ranges are met. This considered a heuristic classification approach [26].

CLIPS provides several approaches to deal with situations when multiple rules will be activated simultaneously and a prioritisation is needed. All rules can be provided with a integer salience value, allowing the rule with the highest salience value to fire first. Alternatively, a CLIPS knowledge-base can be partitioned into thematic modules. The modules are put in a sequence on an execution stack, allowing all rules from the top-most module to fire. Once all rules from this module have been evaluated, processing moves on to the following module [7][8][9].

The rise of the Java programming language led to implementations of languages similar to CLIPS in projects such as JESS, DROOLS and JRules, adopting a similar syntax [2][4]. This family of rule-based and data-driven tools still shares the same basic syntax for the definition of rules to encode human knowledge. While the individual features and capabilities have diverged, it is still possible to port an application if a restricted subset of features is used to write portable programs. As a side effect, a wealth of documentation can be used from these CLIPS-related projects such as g.infer [2][4][6][25].

4. Rete

The core of the Production Rule System CLIPS is an Inference Engine that is able to handle a large number of rules and facts using the Rete algorithm for forward-chaining inference. The word „Rete" is Latin, meaning „net" or „comb". The Rete algorithm was designed by Dr Charles L. Forgie [3]. Rete has become the basis for many popular rule engines and expert system shells apart from CLIPS. It provides a generalized logical description of an implementation of functionality, which is responsible for matching data (facts) against rules (production) in a pattern-matching production system. A production (rule) consists of one or more conditions and a set of actions which may be undertaken for each complete set of facts that match the conditions.

4.1. Pattern-Matching Performance

The efficiency of the Rete pattern-matching algorithm is based on the assumption that data changes slowly over time. This assumption will fail for applications where rapid data change can occur as it is often the case in GIS. Because of this, g.infer should not be perceived as a replacement of optimized GRASS tools such as r.mapcalc, which manipulate each cell of a raster layer [8]. However, in many cases data pre-processing will allow to comply with the Rete assumption. Such approaches include the grouping of data into larger sets, limiting rule-based intervals to elements which transition into another set, or to convert numeric value ranges into symbolic values like "unavailable, nominal, critical", retracting the current fact and assert a new one only if the symbolic value has changed. For g.infer, such pre-processing can be achieved by GRASS modules such as r.clump, r.mask, r.reclass or r.recode.

5. Embedding CLIPS in GRASS

CLIPS and GRASS are based on the C language. Until now, no close-coupling below the API based on the C API has been published. Of the numerous software projects which embed CLIPS in other programming languages, two are currently known to have been used to connect CLIPS and GRASS GIS.

The **CLIPS And Perl with Extensions (CAPE)** project was developed in the late 1990s [11]. CAPE closely integrates CLIPS and the procedural programming language Perl, and provides extensions to facilitate building systems with an intimate mixture of the two [12].

The GRASSCAPE libraries were a merger of CAPE and GRASS5 to provide rule based programming in a GRASS environment [19]. This was used to assess the validity of radar meteorological data products, to generate warning messages for the general public for storm events and to create rainfall intensity maps for soil erosion studies. The development of GRASSCAPE was stopped in 2003.

PyCLIPS is an extension module for the Python language, interfacing it with CLIPS [5]. Python has become increasingly popular for scripting in GRASS GIS, and was selected as the reference language for GRASS 7.0 extensions: A reimplementation and extension of GRASS-CAPE based on PyCLIPS was started in 2011. This development eventually resulted in the GRASS module g.infer.

5.1. The GRASS GIS Module g.infer

g.infer is a GRASS add-on module for GRASS6.4.x and GRASS 7.0. It allows one to define and execute rule-based data-driven processing based on GRASS data layers. Currently, raster-, volume- and point vector-layers can be processed with g.infer. In addition, GRASS environment variables, including the GRASS location region settings can be queried and manipulated within the g.infer Production System.

Access to various parameters of the embedded CLIPS instance is provided by options and flags accessible both from the Command Line Interface or a GRASS Graphical User Interface (GUI). In addition to automated inference runs, an interactive mode allows to interact with the rule-base environment on the fly. The g.infer module further allows access to GRASS modules and their output. This makes the set-up of rule-driven GIS workflows possible. The development of g.infer is currently supported by GISIX.com to develop sample applications and create performance metrics. The module will be released as a GRASS add-on in late 2012.

GIS-based Inference Workflow This section describes the tasks executed by g.infer to allow the processing of GIS-layers by rule-based inference in the CLIPS environment. The involved tasks are best described when using both GRASS- and CLIPS- centered perspectives.

GRASS GIS-centered Workflow From the perspective of a GRASS GIS user, g.infer provides access to a Production Rule System to set-up and maintain specific rule-based data-driven tasks in GRASS as an Expert System. The following steps are required to implement such an Expert System and to conduct a rule-based analysis of spatial data layers:

In the preparation phase, the goal of the analysis must be defined. The knowledge and expertise needed to reach this goal will be written down as plain language rules (Knowledge Engineering). This will likely require the involvement of human domain experts. The plain language rules are in a next step translated into CLIPS rule syntax to be stored in a rule-base file. In this step, the names of the GRASS layers to be queried must be used in the CLIPS rules. The standard naming convention for how to address GRASS layers in CLIPS rules is provided in the g.infer html documentation [21].

In the application phase, g.infer is invoked with the required parameters: The provided GRASS layers and the rulebase-file are ingested. The user can opt to have elements of the rule- or fact-base printed out, or to interact with the inference environment via a command line prompt. Now the inference run will commence unless the early abort flag has not been set. A successful inference run leads to the update of selected GIS data layers, and optionally the creation of new vector layers and log files.

Production System-centered Workflow From the perspective of the CLIPS Production System, it is operating in g.infer in an embedded environment, which provides inference-related parameters and data. The overall CLIPS-based process begins by the setting of Inference Engine-related parameters and the definition of fact-templates for each GIS layer to be used. In turn, the content of the GIS layers is copied into facts for the inference process, and the rule-base is set up from the provided rule-file. At this stage interactive access via a CLIPS prompt can allow one to list and manipulate the current content. If no abort signal is given by the user from the GRASS layer, the inference process starts.

The existing rules are pattern-matched against the existing facts to start the firing of the first rules, resulting in modifications and extensions of the fact-base. This can lead to renewed firing of rules, starting an iterative process. Once the firing of rules ceases, the inference process ends. For the rule-based system, the final updating of GIS layers from parts of the fact-base is transparent.

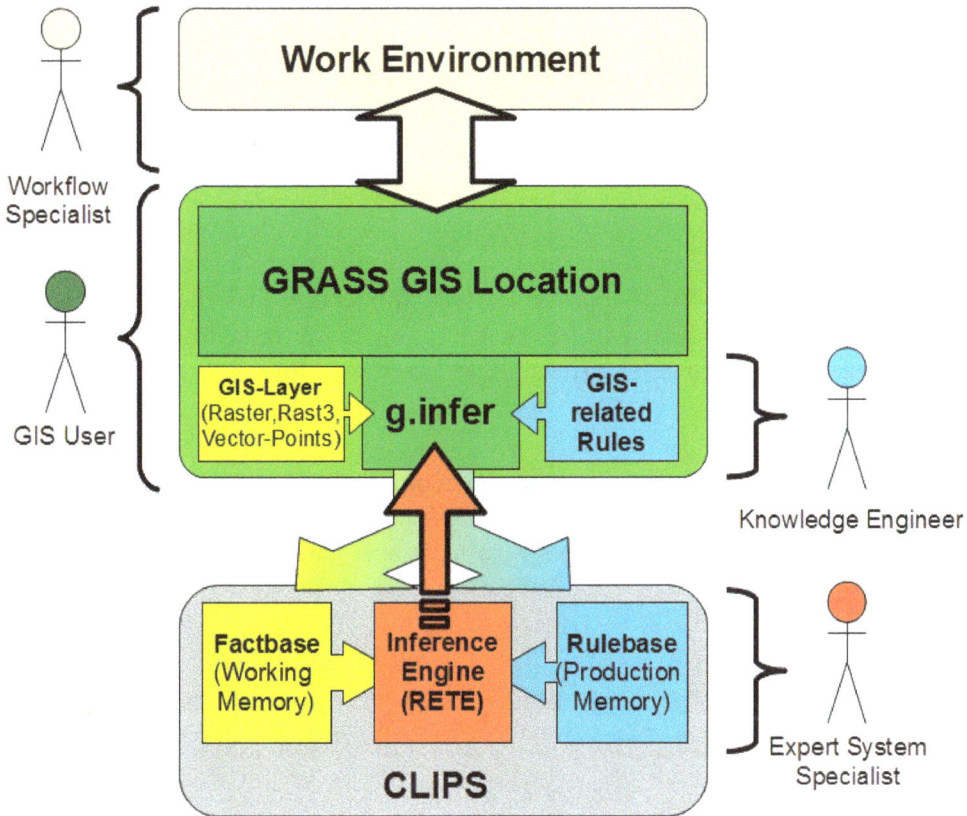

Figure 1: Interaction of the g.infer module with related GRASS GIS components and the CLIPS Production Rule System. Human experts (shown as figurines) can interact with this work environment independently on multiple levels.

Knowledge Engineering and Management   g.infer provides multiple levels of knowledge modelling and rule-programming approaches, allowing one to use increasingly complex techniques when required. The options range from closely-coupled to loosely-coupled inference rules, extensions of the CLIPS language by functions, rulebase stratification by salience values or its partitioning into modules, up to ontologies and object oriented modelling using the COOL language. Depending on the techniques used for knowledge engineering, the evaluation and testing of a rule base can become a complex task. Multiple factors are to be considered, ranging from simple typos to content issues of input data layers and nontrivial rule precedence issues. This requires options for stepwise interactive execution and checking the inference process, the capability to log specific processes for later analysis and to save all or parts of the rule base or the object instances. Such features exist in the CLIPS environment and can be accessed by GIS users as GRASS module flags and -options in g.infer.

6. Business Processes in GRASS GIS

Software projects such a Jess and Drools have been used in the past years to apply the rule-based data-driven approach to new tasks, beyond the classical field of classification topics [2][4].

They are used to set up business-themed systems, in a vertical stack of tasks, including knowledge engineering, management and deployment of rules, collaboration between rule-based systems, analysis and end user tools. Software systems to handle such task stacks are called Business Rule Management Systems.

The emerging methodology of describing the application of rule-based systems in enterprise environments for structured, product-generating activities has been named the Business Rules Approach [2][15]. This is also of interest for GIS application and the related workflow perspective: GRASS GIS modules can be used in a similar fashion to set-up data-processing workflows, while on a higher level, GRASS GIS-based workflows can be fully integrated into greater workflow-chains.

Inference processes in g.infer can trigger the modifications of its fact-base, but they can also be used to execute further GRASS modules or scripted GRASS-workflows. By doing this, it is possible for an inference process to have new data imported into GRASS, have it processed and have the outcome to extend its fact-base. Further, direct queries to the user requesting interactive input can be triggered by rules. This allows one to effectively have a data-processing workflow in GRASS GIS controlled by an inference process within g.infer. The overall process is illustrated in Figure 2. The topic of interaction between COOL object instances and the rule-base will be covered in a later publication.

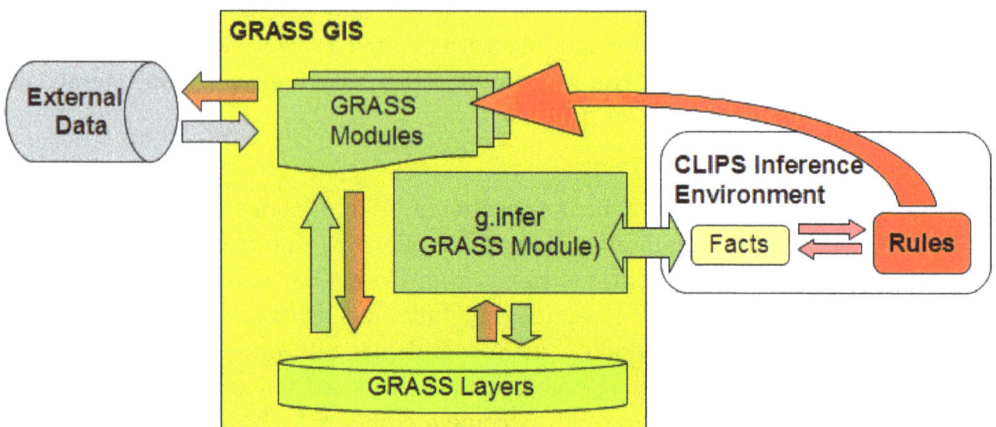

Figure 2: Overview of the interactions between the GRASS GIS environment (green), the CLIPS-based inference environment (blue) and external data sources (grey), highlighting (red) the potential for workflow control to be exerted by the inference process.

7. Application Scenarios

The forerunner modules of g.infer, r.[b]infer and GRASSCAPE have already enabled rule-based inference for GRASS GIS, yet did not succeed to attract a large audience of GIS application developers in the long term. As a consequence, their porting to the current versions of GRASS 6.x and GRASS 7 did not occur for lack of serious need. The same challenge applies for g.infer to give today's community of GRASS application developers significant added value in their work. For this reason, a cookbook-type publication will follow up this introductory article, providing hands on application examples with respective rule bases to lower the learning curve, and to show inference-based applications on varying levels of complexity.

As g.infer connects the domains of GIS, workflows, classification, and rule-based systems, the module can be applied for at least three different application scenarios in general:

1. **GIS-based classification tasks**, where g.infer can be used to quickly set up and apply rule-based data-driven classification of varying levels of complexity. This includes the combination of queries on information from both raster- and vector-layers and derived facts (e.g.: „IF a location is both an archeological site [vector information] and an abandoned mine [raster information] THEN assert geoarcheological monument"). Classification rules can be flexibly chained within the rulebase: „IF a location is an abandoned mine THEN assert no-trespassing"; „IF the location is a monument AND today-is-national monument day THEN assert guided-tours-available"

2. **Workflows and business processes**, where GIS-based data processing chains need to adapt to changing data quality, human user input or time constraints. This allows to use rules similar to the exception-construct known from other programming languages, such as Python: IF "the current GRASS region is smaller than threshold X" THEN "assert no-use-satellite-images; use-aerial-photography".

3. **Extensions for the underlying production system**: In this case, the perception is reversed. From the standpoint of the embedded Production System, the g.infer-provided interface to GRASS GIS is merely an extension for advanced A.I.-focused tasks such as ontology modelling or intelligent software agents: While g.infer rules define knowledge about cause-effect actions among certain entities (facts), an ontology is a structural framework defining the object classes of the entities for the current knowledge domain and their properties and relations (taxonomy). Within g.infer, a simple implicit ontology for the GRASS GIS domain is used, to translate GRASS layers in corresponding domain objects which are defined via CLIPS templates. So the CLIPS templates and their interrelation define the ontology. Any g.infer application can set-up additional ontologies for their specific knowledge domain. Another field for exploration are intelligent rule-based software agents implemented in g.infer. Such autonomous entities are capable to observe and manipulate their surroundings while trying to achieve goals, effectively interacting with the GIS layers by manipulating the corresponding spatially-enabled facts in g.infer. They can be distributed in GIS-geographical space if they posses spatial locations (e.g. virtual sensor networks) and can become mobile if they have also the means to change their current position within the GRASS location. Another option are multi-agent systems (MAS), which are able to communicate to achieve a common goal.

8. Conclusion

The new add-on module g.infer re-introduces generic rule-based data-driven modelling to GRASS GIS for the current versions of GRASS 6.4.x and GRASS 7. It provides a new flexible interface between GRASS GIS-based geoscientific modelling and Artificial Intelligence (A.I.) research based on the C-Language Integrated Production System (CLIPS) toolkit. This allows to develop rule-based data-driven processing of GRASS GIS data sources by Expert Systems encoded as CLIPS knowledge bases.

The description of the theoretical and developmental background of the g.infer module already brings up possible application scenarios, ranging from the rule-driven formulation of hard to describe GIS-classification tasks, the flexible set-up and management of GIS workflows to Artificial Intelligence-focused topics such as the ontology management, defining taxonomies of knowledge domains, and the exploration of intelligent software agents encoded as g.infer rulebases.

Detailed examples of the practical application of g.infer for a range of real world problems will be provided in a follow up publication.

References

[1] Buehler K. (1990). A GIS providing grounds for water resources research

http://docs.lib.purdue.edu/cgi/viewcontent.cgi?article=1188&context=watertech

[2] Buehler K. (1999). r.binfer Documentation

https://svn.osgeo.org/grass/grass/branches/releasebranch_5_5/html/html/r.binfer.html

[3] Browne P. (2009) JBOSS Drools Business Rules. Packt Publishing. ISBN 1-847-19606-3

[4] Forgy C., (1982) Rete: A Fast Algorithm for the Many Pattern/Many Object Pattern Match Problem, Artificial Intelligence, 19

[5] Friedman-Hill E. (2003). Jess in Action. Manning Publications. ISBN 1-930-11089-8

[6] Garosi F. (2008). PyCLIPS Manual Release 1.0.

http://sourceforge.net/projects/pyclips/files/pyclips/pyclips-1.0/pyclips-1.0.7.348.pdf/download

[7] Giarratano J., Gary R. (2004). Expert Systems: Principles and Programming. Course Technology. ISBN 0-534-38447-1

[8] Giarratano, J.C. (2007). CLIPS User's Guide.

http://clipsrules.sourceforge.net/documentation/v630/ug.pdf

[9] Giarratano, J.C. (2007). CLIPS Reference Manual: Basic Programming Guide.

http://clipsrules.sourceforge.net/documentation/v630/bpg.pdf

[10] Giarratano, J.C. (2008). CLIPS Reference Manual: Advanced Programming Guide.

http://clipsrules.sourceforge.net/documentation/v630/apg.pdf

[11] Graham P. (1995). ANSI Common Lisp. Prentice Hall, ISBN 0-133-79875-6

[12] Inder R. (1998) CAPE Users Manual, ETLTechnical Report ETL-TR98-3, Electrotechnical Laboratory, Tsukuba, Japan.

[13] Inder R. (2000) CAPE: Extending CLIPS for the internet, Knowledge-Based Systems 13 (2000), Elsevier

[14] Jasiewicz J. (2011): r.fuzzy GRASS Addons Repository

http://trac.osgeo.org/grass/browser/grass-addons/grass6/raster/r.fuzzy

[15] Jasiewicz J., Di Leo M. (2012): Application of GRASS fuzzy inference system in flood prone areas prediction

http://geoinformatics.fsv.cvut.cz/gwiki/Application_of_GRASS_fuzzy_
inference_system_in_flood_prone_areas_prediction

[16] JBOSS Community Documentation(2008) The Rule Engine

http://docs.jboss.org/drools/release/5.4.0.CR1/drools-expert-docs/html/
ch01.html

[17] Lake M. W. (2000). MAGICAL computer simulation of Mesolithic foraging. In Kohler, T. A. and Gumerman, G. J., editors, *Dynamics in Human and Primate Societies: Agent-Based Modelling of Social and Spatial Processes*. Oxford University Press, New York.

[18] Lake M. W. (2000) MAGICAL computer simulation of Mesolithic foraging on Islay. In Mithen, S. J., editor, *Hunter-Gatherer Landscape Archaeology: The Southern Hebrides Mesolithic Project, 1988-98, volume 2: Archaeological Fieldwork on Colonsay, Computer Modelling, Experimental Archaeology, and Final Interpretations*. The McDonald Institute for Archaeological Research, Cambridge.

[19] Lake M. W. (2002) Magical for GRASS4.x

http://www.ucl.ac.uk/~tcrnmar/simulation/magical/manual/index.html

[20] Löwe P. (2004). Technical Note - A Spatial Decision Support System for Radar-Metereology in South Africa. Transactions in GIS. 8(2), Blackwell Publishing Ltd, Oxford.

[21] Löwe P. (2005). Knowledge Management and GRASS GIS: r.infer, GRASS-Newsletter 01/2005, ISSN 1614-8746

[22] Löwe P. (2012) g.infer Documentation (2012):

http://grasslab.gisix.com/scripts/g.infer/g.infer.html

[23] Lustenberger M (2012) r.agent GRASS Addons Repository

http://trac.osgeo.org/grass/browser/grass-addons/grass7/raster/r.agent

[24] Martin M., Westervelt J. (1991).GRASS4.0 Inference Engine: r.infer

http://grass.osgeo.org/gdp/raster/infer.ps.gz

[25] Netzel P (2011) Implementation of ANN in GRASS – an example of using ANN for spatial interpolation

http://www.wgug.org/images/stories/materialy/20110519praga-ann.pdf

[26] Jackson P. (1998). Introduction to Expert Systems. Addison Wesley. ISBN 0-201-87686-8

[27] Puppe F. (1993). Systematic Introduction to Expert Systems. Springer. ISBN 3-540-56255-9

[28] Riley G., (2008). The History of CLIPS.

http://clipsrules.sourceforge.net/WhatIsCLIPS.html#History

[29] Rudolph G. (2008). Some Guidelines For Deciding Whether To Use a Rule Engine.

http://www.jessrules.com/guidelines.shtml

## 14

Reference Data as a Basis for National Spatial Data Infrastructure

Tomáš Mildorf and Václav Čada

Department of Mathematics - Section of Geomatics
Faculty of Applied Sciences, University of West Bohemia in Pilsen
Univerzitní 22, 306 14 Pilsen, Czech Republic

mildorf@kma.zcu.cz   cada@kma.zcu.cz

Abstract

Spatial data are increasingly being used for a range of applications beyond their traditional uses. Collection of such data and their update constitute a substantial part of the total costs for their maintenance. In order to ensure sustainable development in the area of geographic information systems, efficient data custody and coordination mechanisms for data sharing must be put in place. This paper shows the importance of reference data as a basis for national spatial data infrastructure that serves as a platform for decision making processes in society. There are several European initiatives supporting the wider use of spatial data. An example is the INSPIRE Directive. Its principles and the main world trends in data integration pave the way to successful SDI driven by stakeholders and coordinated by national mapping agencies.

Keywords: reference data, INSPIRE, spatial data infrastructure, data integration

1. Introduction

The role of spatial data in current society is increasingly important. Spatial data help us to shape the environment we live in, to manage the resources we possess and to preserve our cultural heritage. The importance of spatial data is being recognised by decision makers, whose support is essential for further development of spatial information technologies and the wider use of spatial data in practice. Funding schemes at various levels of administration aim to support projects and initiatives dealing with access to heterogeneous spatial data through innovative technologies. The potential of spatial data is also given by the economic value of spatial information within public sector information in the EU. An analysis dated back to 1999 (see Figure 1) is underpinned by recent studies including ACIL Tasman (2008) and Fornefeld et al. (2009). The range of applications where spatial data play an important role is growing alongside the demand for sustainable spatial data management.

Despite the importance of spatial data for society, there are certain questions which need to be addressed in order to achieve sustainable management and efficient use of spatial data. The Geographic Information Panel (2008) declares that "current users of spatial information spend 80 per cent of their time collating and managing the information and only 20 per cent analysing it to solve problems and generate benefits." How can we overcome this imbalance?

Recent activities by the European Commission provided a European Interoperability Framework (EIF) aiming to support the interoperability of European public sector information and

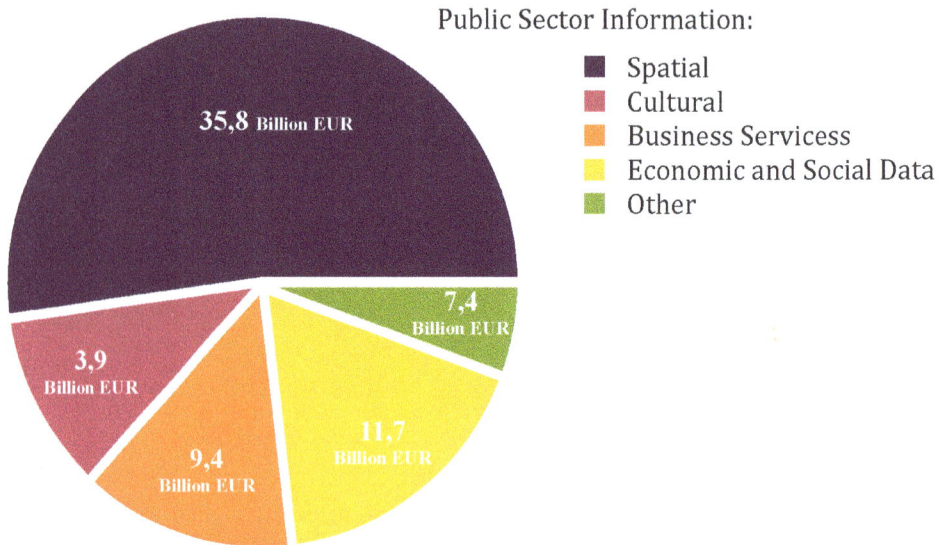

Figure 1: Economic value of public sector information in the European Union in 1999 (Pira International Ltd. & University of East Anglia and KnowledgeView Ltd. 2000).

related services, taking into account legal, organisational, semantic and technical issues. One of the most important projects with a focus on harmonisation of spatial data and services is the Infrastructure for Spatial Information in the European Community (INSPIRE) Directive. INSPIRE entered into force in 2007 as a European directive and over the next two years was transposed into the national legislation of all member states of the European Union. The main objective of the INSPIRE Directive is to establish an infrastructure for spatial information in Europe to "assist policy-making in relation to policies and activities that may have direct or indirect impact on the environment." (European Parliament 2007). Is fulfilling the INSPIRE Directive sufficient to secure the sustainability of spatial data infrastructures (SDI) on a national level? Is there anything that national mapping agencies must beware of?

The spatial data analyses that are needed for decision making processes in society require spatial data of appropriate quality in terms of completeness, logical consistency and positional, temporal and thematic accuracy. One aspect to which insufficient attention is paid in SDI building, and which is considered by the authors as a basis for national SDI, is the delimitation of reference data. What is understood by the authors under the term reference data? What benefits for decision making processes and the sustainability of national SDIs do they present?

In order to address the above mentioned questions the authors analysed selected data sources in the Czech Republic within the context of the INSPIRE Directive. The next chapter reviews the scope of INSPIRE and its main principles. Chapter 3 presents the results of the analysis of the selected data sources from the Czech Republic. Chapter 4 describes the role of reference data within the context of an SDI. The need for reference data is underlined by global trends in data integration and cohesion of cadastral and topographic data in SDI building in Chapter 5. Cases from the Netherlands and Great Britain give the context for the issues tackled in this paper and provide best practice in the national SDI implementation with regard to reference data and maintenance of the INSPIRE principles. The paper aims to start a discussion about these topics.

2. The scope of INSPIRE

The INSPIRE Directive lays down the rules that enable the sharing and reuse of pre-existing data. Heterogeneous spatial data originating from various sources are harmonised according to the common INSPIRE data specifications. A single access point enables users to search the right data for their purposes, to seamlessly view the data and to download them or to perform other spatial services. INSPIRE is a good basis not only for decision makers but also for planners, businesses, emergency management and others.

The success of INSPIRE is based on principles that are crucial for achieving the sustainability of the infrastructure. The INSPIRE principles include:

- Data should be collected once and maintained at the level where this can be done most effectively;

- It should be possible to combine seamlessly spatial data from different sources and share them between many users and applications;

- Spatial data should be collected at one level of government and shared between all levels;

- Spatial data needed for good governance should be available on conditions that are not restricting their extensive use;

- It should be easy to discover which spatial data are available, to evaluate their fitness for purpose and to know which conditions apply for their use. (INSPIRE Website 2012)

All of these principles can be achieved by implementing the INSPIRE mechanisms for data sharing. However, in some cases this is only true to a certain degree. The heterogeneity of spatial data harmonised by INSPIRE can cause certain inconsistencies in the target data, for example when two datasets of different levels of detail are harmonised (see Figure 2). The data provided through the INSPIRE infrastructure for applications requiring data of a high level of detail and high accuracy may not be sufficient. The legislation put in place by the INSPIRE Directive does not affect the collection and processing of data which are considered by the authors as the main sources of inconsistencies. The primary scope of INSPIRE is therefore on data at European, national and regional level where the inconsistencies are diminished by the level of generalisation and the expected level of quality.

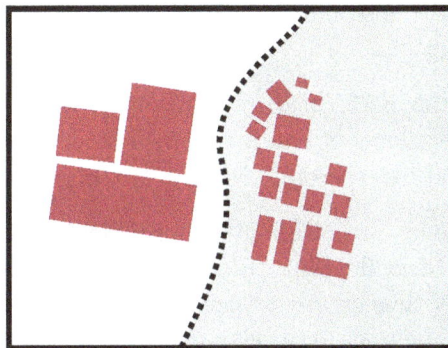

Figure 2: Inconsistency of harmonised datasets of different levels of detail.

INSPIRE represents a solid foundation for the European SDI. National SDIs should benefit from the provisions of INSPIRE. It should be the responsibility of national mapping agencies (NMAs) to combine the national requirements and priorities with INSPIRE and to secure the sustainability of the overall infrastructure. NMAs should take advantage of the INSPIRE implementation and trigger the creation of national SDI.

3. Analysis of selected data sources in the Czech Republic

The authors analysed the situation of spatial data management in public administration in the Czech Republic in relation to the INSPIRE Directive and its principles. The focus was mainly on semantic aspects of selected digital data sources of higher level of detail including:

- Cadastral map (KM);

- Technical maps (TMO);

- Planning Analytical Materials (UAP);

- Fundamental Base of Geographic Data (ZABAGED).

All the geographic features from these data sources were compared in terms of their definition and an overview of relations between the features was drawn. Due to complexity of the overview including the definitions of all the features only an indicative table showing the similarities between the analysed data sources is presented (see Table 1).

The results of this analysis show that many geographic features from the selected data sources are duplicated, they are not maintained at the most appropriate level and it is not easy to combine them with other data sources. Three of five INSPIRE principles are not maintained and the sustainability of the data sources forming the national SDI is questionable.

4. Reference data

Data collection and their update make a substantial part of the total costs of data maintenance. Sharing of spatial data between different applications enables sharing of the costs for data management. The historical development of spatial data of public administration in many countries, and the lack of coordination between data producers and data users, has led to duplication in data collection. An example is the situation in the Czech Republic analysed by the authors. Real world phenomena are independently captured, processed, stored and updated by several organisations.

Based on the findings of the performed analysis, the authors propose the delimitation of reference data on the highest level of detail that will be shared between several applications of public administration and other users.

Great achievement was done by establishing the basic registers in the Czech Republic including the Register of Territorial Identification, Addresses and Real Estates (RUIAN). Geographic features included in RUIAN thus create a reference base for all applications of public administration and other users. The system of basic registers uses the term reference data as data maintained in basic registers and given by law that are up-to-date, valid and unambiguous for every application of public administration.

DKM	ZABAGED	TMO	UAP
hrana koruny a střední dělicí pás silnice nebo dálnice	silnice, dálnice		dálnice včetně ochranného pásma rychlostní silnice včetně ochranného pásma silnice I. třídy včetně ochranného pásma silnice II. třídy včetně ochranného pásma silnice III. třídy včetně ochranného pásma
osa kolejí železniční tratě mimo železniční stanici a průmyslové závody	železniční trať	osa železničních a tramvajových kolejí	železniční dráha celostátní včetně ochranného pásma dráhy celostátní vybudované pro rychlost větší než 160 km/h včetně ochranného pásma
	lanová dráha, lyžařský vlek stožár lanové dráhy	lanová dráha, dopravník	lanová dráha včetně ochranného pásma
nadzemní vedení vysokého a velmi vysokého napětí včetně stožárů	elektrické vedení	elektrické vedení	nadzemní a podzemní vedení elektrizační soustavy včetně ochranného pásma
stavba k využití vodní energie (vodní elektrárna)	elektrárna	elektrárna, spínací stanice nebo měnírna, transformovna, transformační stanice	výrobna elektřiny včetně ochranného pásma
	dálkový produktovod	potrubí produktovodu	produktovod včetně ochranného pásma
	vodojem věžový	vodojem	
ostatní plocha - skládka	skládka		skládka včetně ochranného pásma
stavba odkaliště	usazovací nádrž, odkaliště	odkalovací nádrž, kaliště	odval, výsypka, odkaliště, halda
vodní plocha - rybník	vodní plocha	vodní nádrž	vodní útvar povrchových, podzemních vod
vodní plocha - vodní nádrž přírodní			
vodní plocha - vodní nádrž umělá			
vodní plocha - koryto vodního toku přirozené nebo upravené	vodní tok	vodní tok	
vodní plocha - koryto vodního toku umělé	břehová čára	vodní tok občasný, vysychající, odpadová stoka, suchý příkop	
vodní plocha - zamokřená plocha	bažina, močál	močál	
národní park - I. zóna	velkoplošné zvláště chráněné území	hranice zvláště chráněného území	národní park včetně zón a ochranného pásma
národní park - II. zóna			
národní park - III. zóna			
ochranné pásmo národního parku			
chráněná krajinná oblast - I. zóna			
chráněná krajinná oblast - II.-IV. zóna		hranice ochranného pásma	chráněná krajinná oblast včetně zón
pam. rezervace - budova, pozemek v památkové rezervaci		hranice památkové rezervace	památková rezervace včetně ochranného pásma
pam. zóna - budova, pozemek v památkové zóně		hranice památkové zóny	památková zóna včetně ochranného pásma
chr.lož.území,dob.prostor,chr.území pro zvl.zásahy do z.kůry		hranice chráněného ložiskového území	chráněné ložiskové území

Table 1: The comparison of geographic features from the selected data sources.

Reference data were used as the basis for documents forming the current INSPIRE Directive. The Chapter on Reference Data of the European Territorial Management Information Infrastructure (ETeMII) White Paper (2001) defined the following functional requirements for reference data:

- to provide an unambiguous location for a user's information;

- to enable the merging of data from various sources;

- to provide a context to allow others to better understand the information that is being presented. (p. 5)

The INSPIRE TWG Cadastral Parcels (2009) defines reference data as data that constitute the spatial frame for linking and pointing at other information that belongs to specific thematic fields; e.g. land use, land cover, agriculture and demography. These are considered as application data, which is a complementary term to reference data (INSPIRE Drafting

Team Data Specifications, 2008). Reference data provide a common link between various applications and provide mechanisms for sharing information in society. In the initial phase of the INSPIRE development the following spatial data themes were defined as reference data (RDM Working Group 2002):

- geodetic reference system;
- units of administration;
- units of property rights;
- addresses;
- selected topographic themes;
- orthoimagery;
- geographical names. (p. 11)

It is necessary to note that these are spatial data themes covering a wide range of geographic features. The vision of the authors goes a step beyond the spatial data themes into a selection of particular geographic features that fulfil the above mentioned functional requirements for reference data. The selection of geographic features playing the role of reference data should be available for any application. The consensus on the features, their quality and the sources of updates as well as the forms of exchange should be agreed.

The next chapter introduces examples from Great Britain, the Netherlands and other countries to give context to the above mentioned ideas, especially in relation to reference data and their integration with application data.

5. Trends in data integration

5.1. Great Britain

Great Britain realised the importance of INSPIRE in time and in 2008 the UK Location Programme[1] was created. The programme combines the national priorities (implementation of the UK Location Strategy) with the requirements of INSPIRE and provides a complex solution for the sharing and reuse of spatial information of public administration.

One of the operational challenges of the UK Location Programme is, according to the Geographic Information Panel (2008), to develop a set of core reference geographies captured at the highest level of detail. In the initial stage, the core reference geographies should include a geodetic framework, topographic mapping (at different resolutions and including ground height information), geographic names, addresses, streets, land and property ownership, hydrology/hydrography, statistical boundaries and administrative boundaries.

One of the main building blocks of the UK Location Programme is the Digital National Framework[2] (DNF) that includes the interoperability components such as feature catalogue, terminology, metadata, standards and reference model. In relation to reference data, the feature catalogue of the DNF Base Reference Objects is of most importance and provides an

---

[1] http://location.defra.gov.uk/programme/
[2] http://www.dnf.org/

agreed definition and a set of attributes for every geographic feature. The feature catalogue is not definite and can be further extended.

The basis for the current feature catalogues represents the feature catalogue of the Ordnance Survey (OS) MasterMap. OS MasterMap is a common reference base containing a variety of information in four different product layers: Address Layer, Imagery Layer, Integrated Transport Network Layer and Topography Layer. The OS MasterMap database contains over 450 million geographic features. "Every feature within the OS MasterMap database has a unique common reference (a TOID®) which enables the layers to be used together, including the layer of your own information." (Ordnance Survey 2012). The example of the OS MasterMap Topography Layer is depicted in Figure 3.

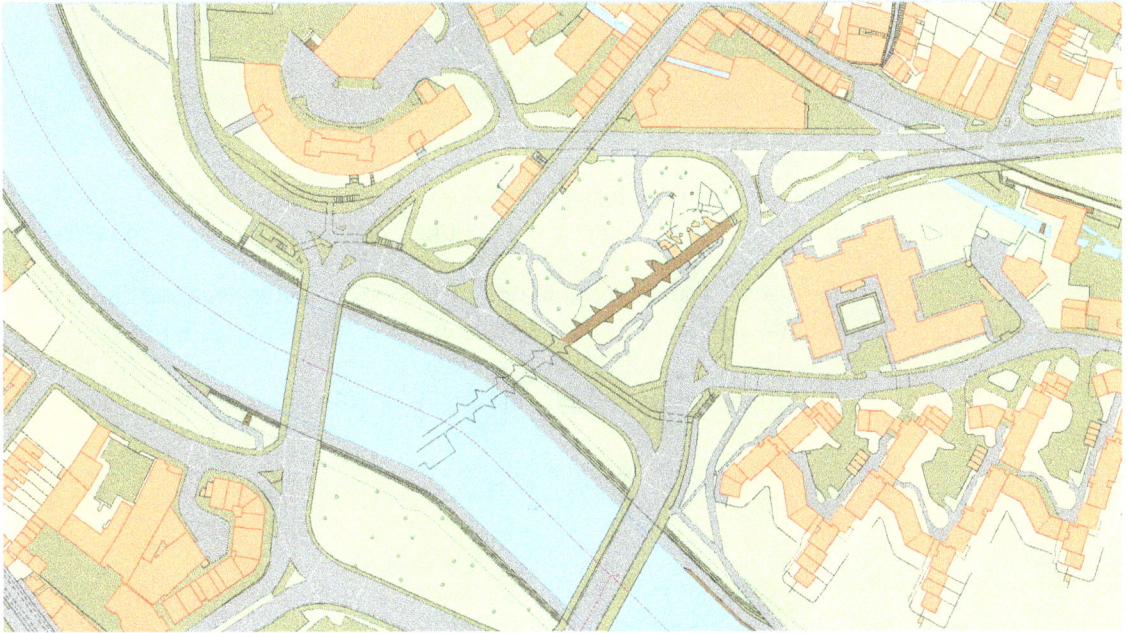

Figure 3: OS Master Map – Topography Layer (Ordnance Survey 2012)

The use of the common reference data aims mainly to improve interoperability, data harmonisation and spatial data quality and increase cross-sector collaboration. The end benefits of reference data encompass (Jones & Wilks 2012):

- reduction of costs for public, private and 3rd sector users of data;

- improvement of quality, efficiency and delivery of services;

- improvement of evidence base for informed policy development and decision making;

- increase of research, innovation and commercial exploitation of location data to benefit UK economy;

- facilitation of other Government initiatives using location based information and tools.

5.2. The Netherlands

The strategic features of the national SDI in the Netherlands are the key registers (*basisregistratie*) that have been developing since 2008. The system of key registers is coordinated by the Ministry of the Interior and Kingdom Relations. There are 13 key registers planned. 4 of them related to spatial information and maintained by Kadaster, the Dutch Land Registry Office, include:

- Key Register of Topography (*Basisregistratie Topografie*, BRT) – contains the topographic dataset Topo10NL in the level of detail equivalent to scale 1:10 000;

- Key Register of Cadastre (*Basisregistratie Kadaster*, BRK);

- Key Register of Large-scale Topography (*Basisregistratie Grootschalige Topografie*, BGT)

- Key Register of Addresses and Inhabitants (*Basisregistraties Adressen en Gebouwen*, BAG).

The key registers provide reference data for applications of public administrations as well as for private sector. The use of common reference data is obligatory for all public administrations.

The inclusion of the large-scale topography as one of the key registers enables linking non-spatial and spatial information of public administration. Non-spatial data can be analysed in the context of spatial data, e.g. visualisation of average income or violence in certain areas (Peersmann et al. 2009). The current and potential uses of this key register are depicted in Figure 4.

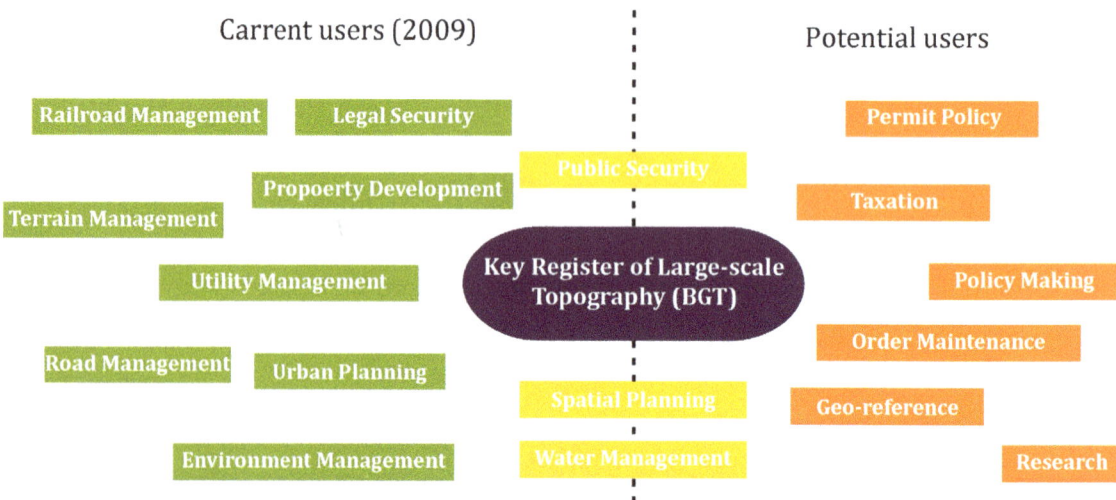

Figure 4: Current and potential uses of BGT (adapted from Peersmann et al., 2009).

An important feature of the Dutch SDI is that cadastral information is maintained together with the large-scale topography in one database. According to Steudler et al. (2009), the layer of buildings is shared between these datasets and cadastral boundaries are aligned to the topographic features.

5.3. Other examples

During the international conference 'Spatial Information for Sustainable Development' in Nairobi, Ryttersgaard (2001) presented experiences from SDI implementation. One of his visions for further development in this area stated: "Cadastral, topographic and thematic datasets should adopt the same overarching philosophy and data model to achieve multi-purpose data integration, both vertically and horizontally." (Ryttersgaard 2001, p.7).

One of the main priorities of the Working Group 2 of the Permanent Committee on GIS Infrastructure for Asia and the Pacific is the integration of datasets containing representation of man-made objects and natural objects. This is performed mainly by integrating cadastral and topographical data which are in most cases maintained separately with no existing links between them and which hinder further exploitation (Rajabifard & Williamson 2006).

The National Land Survey of Finland started with providing free access to reference data since May 2012 including topographic maps, imagery and digital elevation model. The publication aims to enable combination of application data from various sectors with common reference data (Ratia 2012). According to Koski (2011), economic growth should be stimulated.

The interoperability of spatial data in Germany is coordinated by the Working Committee of the Surveying Authorities of the German Länder (AdV). The key feature of the interoperability is the AAA concept for modelling of spatial information based on ISO and OGC standards (Working Committee of the Surveying Authorities of the German Länder 2011). AAA stands for AFIS-ALKIS-ATKIS which represent the geodetic control, cadastral and topographical systems of Germany.

6. Conclusions

The requirements of the users at a local level go beyond the scope of INSPIRE, especially in terms of data quality. National SDIs supported by NMAs should serve as a basis for INSPIRE. Based on the analysis of the situation in the Czech Republic, we can state that pure implementation of the INSPIRE mechanisms for data sharing without introducing the national context is not sufficient for sustainable national SDI building. NMAs should take the responsibility for coordination of the legal, organisational, technical and semantic aspects; meeting the national priorities and user requirements at local, regional and national level; and supporting the implementation of INSPIRE in connection with a national SDI.

The building of a national SDI should use the experience of already existing infrastructures and best practices in data integration. Several examples were mentioned in the paper in order to provide the underlying information for the authors' suggested approach for designing and implementing the national SDI in the Czech Republic.

The trends in data integration aim to combine cadastral and topographic data and to provide a reference data for various applications. The INSPIRE principles are taken as key priorities in SDI building. Based on the performed analyses and long-term research activities, the authors are proposing to define reference data at the highest level of detail, so that they can be shared between various organisations of public administration and the private sector. The delimitation of reference data is expected on the level of geographic features. The understanding of the semantics of available data, together with organisational and legal aspects, is essential for further progress in national SDI building. The Czech NMA should coordinate this process in connection with lead experts from the field of geomatics and geoinformatics. A good progress in this matter started with the establishment of the Register of Territorial Identification, Addresses and Real Estates (RUIAN). However, wider scope and inclusion of other data sources is necessary for better exploitation of spatial data available in public administration and for securing the sustainability of data management.

The definition of reference data mainly represents, but is not limited to, the following benefits:

- sustainable development of the national SDI and support of its extensive use;
- saving costs for data collection and data update of public administration;
- source of guaranteed high quality data;
- unique opportunity for integration of cadastral and topographic information;
- possibility of integration with application data;
- general usability and data availability.

As already mentioned in the introduction, the aim of the paper is to raise a discussion about these topics. What is your opinion?

References

[1] ACIL Tasman, 2008. *The Value of Spatial Information*, Available at: `http://www.crcsi.com.au/Documents/ACILTasmanReport_full.aspx` [Accessed May 31, 2012].

[2] INSPIRE Website, 2012. Available at: `http://inspire.ec.europa.eu/`.

[3] European Parliament, 2007. DIRECTIVE 2007/2/EC OF THE EUROPEAN PARLIA-MENT AND OF THE COUNCIL of 14 March 2007 establishing an Infrastructure for Spatial Information in the European Community (INSPIRE). Available at: `http://eurlex.europa.eu/JOHtml.do?uri=OJ:L:2007:108:SOM:EN:HTML` [Accessed May 31, 2012].

[4] European Territorial Management Information Infrastructure, 2001. ETeMII White Paper: Chapter on Reference Data. Available at: `http://www.ec-gis.org/etemii/reports/chapter1.pdf` [Accessed March 30, 2012].

[5] Fornefeld, M. et al., 2009. *Assessment of the Re-use of Public Sector Information (PSI) in the Geographical Information, Meteorological Information and Legal Information Sectors*, Düsseldorf, Germany: MICUS Management Consulting GmbH.

[6] Geographic Information Panel, 2008. *Place matters: the Location Strategy for the United Kingdom*, Great Britain.

[7] INSPIRE Drafting Team Data Specifications, 2008. *D2.5: Generic Conceptual Model, Version 3.0*, Available at: `http://inspire.jrc.ec.europa.eu/reports/ImplementingRules/DataSpecifications/D2.5_v3.0.pdf` [Accessed May 31, 2012].

[8] INSPIRE TWG Cadastral Parcels, 2009. D2.8.I.6 INSPIRE Data Specification on Cadastral Parcels – Guidelines, v3.0.

[9] Jones, G. & Wilks, P., 2012. UK Location Programme, Benefits Realisation Strategy. Available at: `http://data.gov.uk/sites/default/files/Benefits%20Realisation%20Strategy%20v2.0%20Final.pdf`.

[10] Koski, H., 2011. Does Marginal Cost Pricing of Public Sector Information Spur Firm Growth?

[11] Ordnance Survey, 2012. OS MasterMap - definitive geographical information of Britain. Available at: `http://www.ordnancesurvey.co.uk/oswebsite/products/os-mastermap/index.html` [Accessed April 17, 2012].

[12] Peersmann, M., Eekelen, H. & Meijer, M., 2009. The Large Scale Topographic Base Map of the Netherlands (GBKN): The Transition from a Public-Private Partnership (PPP) to a Legally Mandated Key Registry (BGT). In GSDI World Conference. Rotterdam, The Netherlands. Available at: www.gsdi.org/gsdiconf/gsdi11/papers/pdf/267.pdf [Accessed March 21, 2012].

[13] Pira International Ltd. & University of East Anglia and KnowledgeView Ltd., 2000. *Commercial exploitation of Europe's public sector information*, Luxembourg: Office for Official Publications of the European Communities. Available at: `http://www.ec-gis.org/docs/F15363/PIRA.PDF` [Accessed August 3, 2011].

[14] Rajabifard, A. & Williamson, I., 2006. Integration of Built and Natural Environmental Datasets within National SDI Initiatives. In Seventeenth United Nations Regional Cartographic Conference for Asia and the Pacific. Bangkok, Thailand: United Nations.

[15] Ratia, J., 2012. SDI Interviews Jarmo Ratia of National Land Survey of Finland on Open Data. Available at: `http://www.sdimag.com/20120302584/SDI-Interviews-Jarmo-Ratia-of-National-Land-Survey-of-Finland-on-Open-Data.html` [Accessed May 31, 2012].

[16] RDM Working Group, 2002. Reference Data and Metadata Position Paper. Available at: `http://inspire.jrc.ec.europa.eu/reports/position_papers/inspire_rdm_pp_v4_3_en.pdf` [Accessed May 31, 2012].

[17] Ryttersgaard, J., 2001. SPATIAL DATA INFRASTRUCTURE, Experiences and Visions.

[18] Steudler, D. et al., 2009. Cadastral Template, A Worldwide Comparison of Cadastral Systems. *Cadastral Template, A Worldwide Comparison of Cadastral Systems.* Available at: `http://www.fig.net/cadastraltemplate/index.htm` [Accessed May 31, 2012].

[19] Working Committee of the Surveying Authorities of the German Länder, 2011. *National Report 2010/2011*, Working Committee of the Surveying Authorities of the German Länder. Available at: `http://www.adv-online.de` [Accessed May 31, 2012].

## 15

PostGIS-Based Heterogeneous Sensor Database Framework for the Sensor Observation Service

Ikechukwu Maduako

Institute of Geoinformatics, University of Münster, Germany

Director of Studies, Center for Advanced Spatial Technologies & Mapping (CAST-MP)

Abuja, Nigeria

iykemadu84@gmail.com

Abstract

Environmental monitoring and management systems in most cases deal with models and spatial analytics that involve the integration of in-situ and remote sensor observations. In-situ sensor observations and those gathered by remote sensors are usually provided by different databases and services in real-time dynamic services such as the Geo-Web Services. Thus, data have to be pulled from different databases and transferred over the network before they are fused and processed on the service middleware. This process is very massive and unnecessary communication and work load on the service. Massive work load in large raster downloads from flat-file raster data sources each time a request is made and huge integration and geo-processing work load on the service middleware which could actually be better leveraged at the database level. In this paper, we propose and present a heterogeneous sensor database framework or model for integration, geo-processing and spatial analysis of remote and in-situ sensor observations at the database level. And how this can be integrated in the Sensor Observation Service, SOS to reduce communication and massive workload on the Geospatial Web Services and as well make query request from the user end a lot more flexible.

Keywords: Heterogeneous Sensor Database, PostGIS 2.0, Sensor Observation Service.

1. Introduction

Geo-sensors gathering data to the geospatial sensor web can be classified into remote sensors and in-situ sensors. Remote sensors include satellite sensors, UAV, LIDAR, Aerial Digital Sensors (ADS) and so on, measuring environmental phenomena remotely. These sensors acquire data in raster format at larger scales and extent. In-situ sensors are spatially distributed sensors over a region used to monitor and observe environmental conditions such as temperature, sound intensity, pressure, pollution, vibration, motion etc. These sensors are measuring phenomena in their direct environment and could be said to acquire data in vector data format.

Most environmental monitoring and management systems combine these diverse datasets from heterogeneous sensors for environmental modeling and analysis. For example in monitoring of crop Actual Evapotranspiration at some locations in most cases involves aggregation of remote and in-situ sensor observations [1]. Remote and in-situ sensor data aggregation for real-time calculation of daily crop Gross Primary Productivity GPP such as implemented in a dynamic web mapping service for vegetation productivity [2] and in the marine information system [3] are good examples too.

Meanwhile the process of fusing and processing of these sensor data on the web service currently involves massive data retrieval from different sensor databases, most especially from the raster databases, geo-processing and spatial analytics on service middleware. For web services, this is massive work and communication load over the network and on the service. A sensor database management framework combining remote and in-situ observations would be of great impact to environmental monitoring and management systems. Having these disparate sensor data on one database schema can be leveraged in the geospatial web services to reduce excessive work load and data transfer through the network. Most of the data fusion, aggregations and processing done by web services can be carried out at the database backend and the results delivered to the client through the appropriate web services.

Figure 1 is a diagrammatical illustration of our proposed approach, whereby in-situ and remote sensor data are passed to the proposed heterogeneous sensor database. Data integration and processing are carried out at the database level within the SOS and geo-scientific query or request results are delivered to the clients through the service.

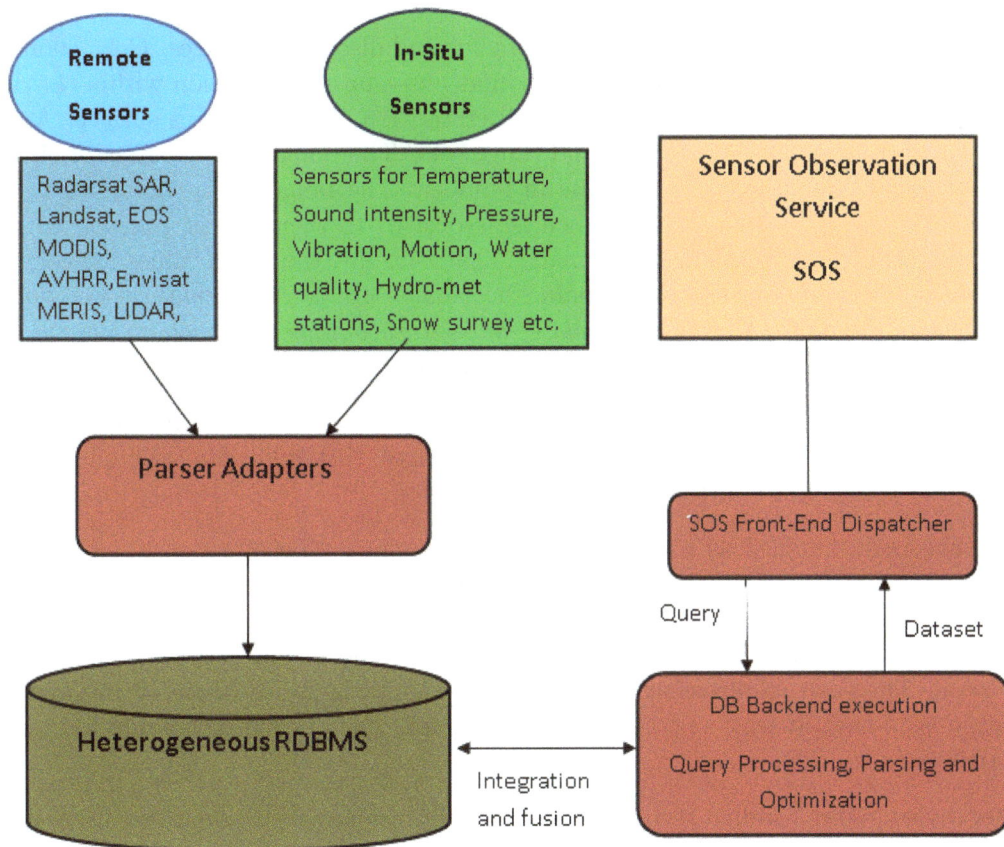

Figure 1: The Conceptual Diagram

2. Requirement Analysis

Firstly we had to analyse the fundamental conceptual and practical requirements for the proposed heterogeneous sensor database framework for a seamless integration of remote and in-situ sensor observations at the database level of the SOS. The analysis is done taking into consideration the varying properties and the underlying structure or format of these two different sensor datasets (raster and vector). The database model for this purpose can be design as a spatial database model based on the Open Geospatial Consortium, OGC standards. Adopting the coverage concept, sensor observations can be approached in coverage perspective. That is to say we can treat in-situ sensor observations as time series vector coverage and remote sensor observations as also time dependent raster coverage.

Coverages have some fundamental properties, exploring some of these properties and how vector and raster coverages inherit these properties, we can conceptually map out an intersection that will underline the seamless integration of vector and raster (in-situ and remote sensors) coverages in a heterogeneous sensor database schema, see figure 2.

According to ISO 19123: 2005 *"a coverage domain consists of a collection of direct positions in a coordinate space that may be defined in terms of up to three spatial dimensions as well as a temporal dimension"* [4].

A coverage is created as soon as a way to query for a certain value given a location is created.

Coverages can be categorised into two, continuous and discrete coverages. Continuous coverage returns a different value of a phenomenon at every possible location within the domain. Discrete coverages can be derived from the discretisation of a continuous surface. A discrete coverage consists of different domain and range sets. The domain set consists of either spatial or temporal geometry objects, finite in number. The range set is comprised of a finite number of attribute values each of which is associated to every direct position within any single spatio-temporal object in the domain. That is to say, the range values are constant on each spatio-temporal object in the domain. *"Coverages are like mathematical functions, they can calculate, lookup, intersect, interpolate, and return one or more values given a location and/or time. They can be defined everywhere for all values or only in certain places "* [5].

Raster and vector coverages are both types of discrete coverage. They differ only in how they store and manage their collection of data. As coverages, they allow for basic query functions such as select, find, list etc. to be carried out on them.

Vector coverages handled as tables are the most common type of coverage implemented in most of the spatial database management systems. Individual data item are stored on each row in the table. The columns of the table ensure that collection is self-consistent. Texts are placed in text columns, numbers in numeric columns, and geometries in geometry columns and so on. The basic requirement a table must have for potential supply of information to a coverage is to have at least one geometry column and one additional column for a value or an attribute.

Raster coverages are handled as an array of multidimensional discrete data as discussed in [6]. In PostgreSQL/PostGIS 2.0 [7] precisely they are stored as regularly gridded data with the geometry of the domain as points and the range could be one or more numeric values (for example number of bands). Text values and timestamps may not be possible.

Hence, in-situ observations (vector data) can be stored in tables with rows and columns in a relational manner, having one-to-one or one-to-many relationship. On the other hand remote sensor observations (raster) cannot reasonably be stored in tables but as gridded multidimensional array of data (array of points). That is to say we only have to leverage the concept of coverage to integrate the two tables in the database. The possible common column for the two datasets (tables) is the geometry column.

Therefore the fundamental requirement from this analysis that could enable us to integrate remote and in-situ sensor observations in a common database could be outlined as:

- storage of in-situ observations as vector point coverage and

- storage remote sensor observations as raster point (pixel) coverage.

Figure 2 is the UML (Unified Modeling Language) model of the concept and management of coverages describing features, relationships, functions and how they present in the database. The insight to this UML model was extracted from the coverage concept model discussed in PostGIS Wiki [5].

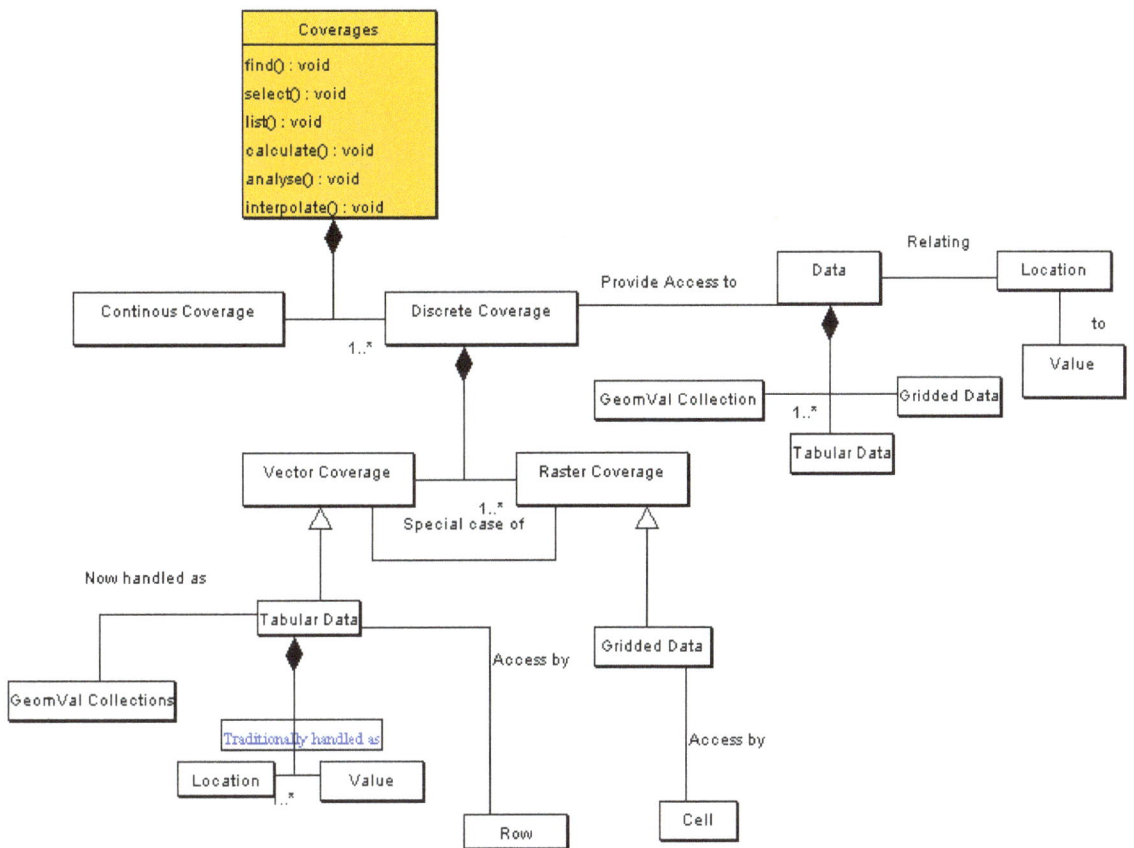

Figure 2: UML Conceptual Model of the concept and Management of Coverages

Leveraging these functions and operations that can be carried out on coverages, the database can offer fundamental operations and functions such as intersection, buffering, overlay, interpolation etc for geo-scientific analysis and processing involving in-situ and remote sensor observations (vector and raster coverages).

With these operations, we can easily run queries for example that can lift a point on the vector coverage, intersect it with the geometrically corresponding point or cell on the raster coverage on the database and return a value. The goal is for us to be able to do relation and overlay operations on the different coverages irrespective of how the coverages are stored. Therefore we need a database management system that can provide these supports for this purpose.

Database Management System (DBMS) Support Analysis

Effective storage and retrieval of vector data has been well developed and implemented in most of the spatial databases such as Postgresql/PostGIS, Oracle Spatial, MySQL, Microsoft SQL Server 2008, SpatiaLite, Informix, etc. On the other hand, Oracle Spatial and Postgresql/PostGIS DBMS are currently the only DBMS that have substantial support for raster data management. Meanwhile Oracle Spatial supports raster data storage with less support for raster data analysis in the database. However PostgreSQL/PostGIS 2.0 has relatively good raster support, functions and operations that we can leverage for the feasibility of our research goal. In addition PostgreSQL/PostGIS 2.0 can be configured with python GDAL-bonded to leverage more functionality.

PostgreSQL/PostGIS 2.0 capability to carry out seamless vector and raster data integration makes it favourable in this type of our work than Oracle Spatial. PostgreSQL/PostGIS 2.0 can handle pixel-level raster analysis unlike Oracle Spatial whose content search is based on Minimum Bounding Rectangle (MBR). PostgreSQL/PostGIS 2.0 uses Geospatial Data Abstraction Libraries (GDAL) to handle multi-format image input and output and when working with out-db-raster, this is a powerful functionality.

PostgreSQL/PostGIS 2.0 supports *GiST spatial indexing, GiST stands for "Generalized Search Tree" and is a generic form of indexing. GiST is used to speed up searches on all kinds of irregular data structures (integer arrays, spectral data, etc) which are not amenable to normal B-Tree indexing* [8].

In PostgreSQL/PostGIS 2.0, raster coverage can be created by having a geometry column called raster and attribute columns containing the attributes to the raster (e.g. band number, timestamp and so on). The fundamental database or storage support needed on the raster coverage for efficient seamless integration and analysis with vector coverages such as tiled raster storage, georeferencing, multiband/multi-resolution support and so on are provided by PostgreSQL/PostGIS 2.0 [9]. Structured Query Language (SQL) raster functions and operators for raster manipulations and analysis are substantially supported in PostgreSQL/PostGIS 2.0, more functions are being developed.

3. Conceptual Design and Modelling of a Heterogeneous Database Schema

Based on those fundamental requirement analysis, we went on to develop a conceptual model of a heterogenous sensor datbase schema, integrating remote and in-situ sensor observations. The UML model in figure 3 shows the high level abstraction model of the fundamental classes (tables) that are needed in a sensor database, their attributes and important operations that can be carried out on them. It shows the relationships and the logic between the classes which enable integration between the classes. The Entity Relationship (ER) diagram in figure 4 describes the logical design for physical implementation of the entities, the fields in each class and the relationships between entities. Also in this section we developed the conceptual model of how the database model can be integrated with other web services seamlessly, introducing the concept of the Web Query Service, WQS.

3.1. The Heterogeneous Database Schema Entity Description

This section describes the functions and relationships of the entities or tables in the heterogeneous sensor database schema as modeled in the UML diagram shown in figure 3. The detailed description of their attributes and values are not necessary within the context of this paper. Figure 3 is the high level conceptual model while figure 4 is the logical model of the database.

The **List_of_Table** class contains the list of all the table names in the database. Operations like GetList_of_Tables and UpdateList_of_Table can be perform on it from the user end through our proposed SQL Web Query Service WQS. The efficacy of this table is to present to the user the names and descriptions of all the tables contained in the database. A "`select * from list_of_table`" SQL instruction from the client end would present a table describing all the tables contained in the database. This is a kind of DescibeTables operation by the user from the client end.

Coverage class holds the information about each of the coverages contained in the database such as the id, description etc. The in-situ and remote sensor observations are stored as coverages, vector and raster respectively in the database. Therefore it is necessary to have a table that presents the collections and a short description of the coverages contained in the database.

Observed_Phenomenon class is the table that contains the names, descriptions, coverage type etc. of the various geographic phenomena that are contained in the observations. This is different from the features of interest table which contains the different features or formats of these observed phenomena that are of special interest.

Feature_of_Interest class is the table that has the records of different features of the observed geographic phenomenon in the database.

SensorPlatform class is the table with the record of the sensor platforms on which the sensors are mounted or housed.

Sensor class is the table that contains the basic attributes about the observing sensor. Attributes such as the sensor platform, sensor type, sensor model etc. are contained in this table.

SensorInfo table contains information related to the sensor mearsurement and method. Attributes such as spatial coverage, temporal coverage, collection frequency, unit of measurement etc. can be found in this table.

Observation class is the table that connects the Sensor, Observed_Phenomenon, Quality, In-situObservations and RemoteObservations tables. Observation table does not contain the values and time stamps of each observed value, they are contained in the in-situ and remote observations tables.

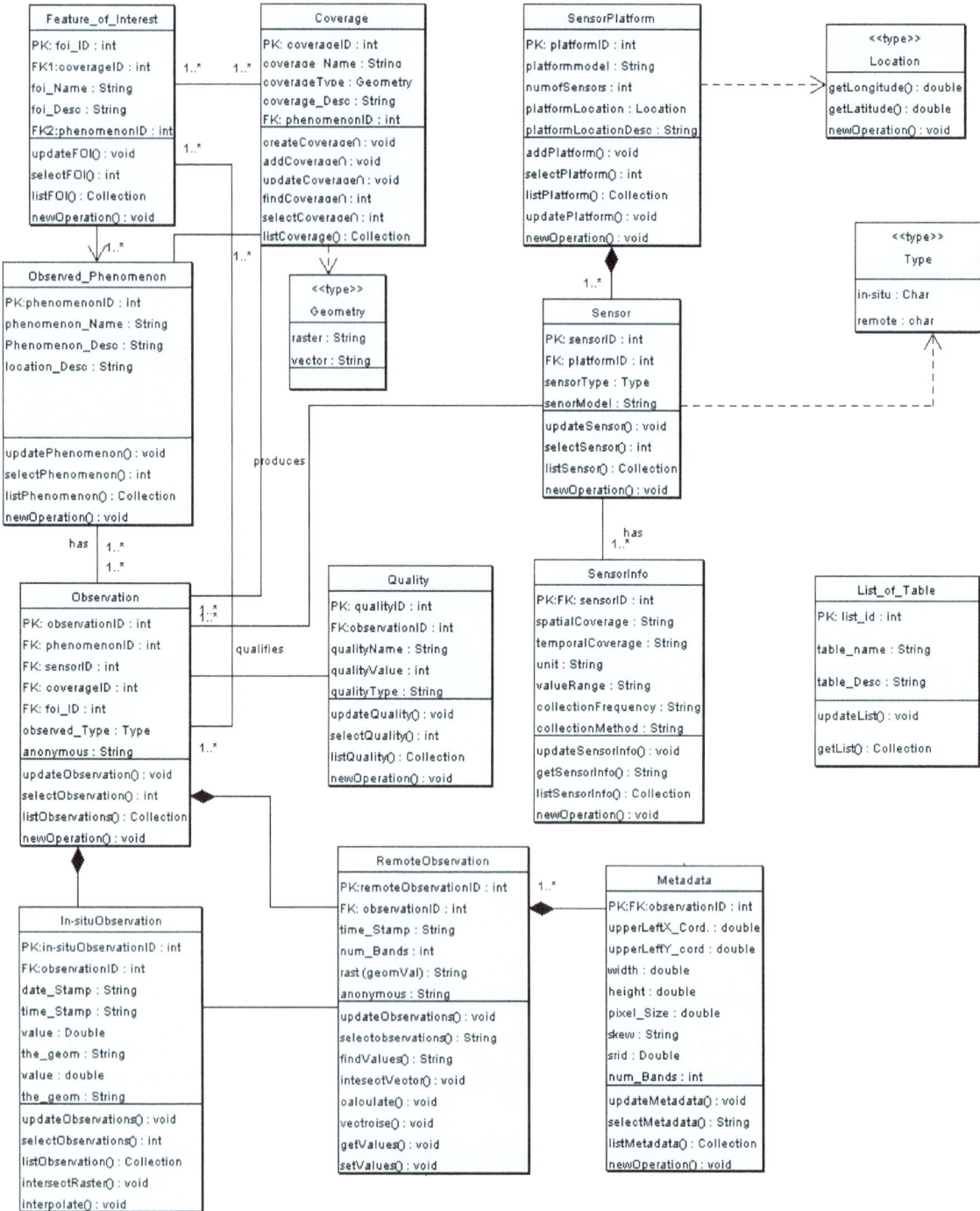

Figure 3: UML Conceptual Schema Model of the Proposed Heterogeneous Sensor Database

In-situObservation class is the table that contains the compelete data of each observation that is contained in the Observation table where observationType is in-situ. It has one-to-one relationship with Observations table. The relationship between this table and the remote_observation table are handled on the fly leveraging the PostGIS intersection operation because the two coverages are handled differently in the database. Basic operations as well as complex operations such as intersection with raster, interoplation or rasterisastion can be carried out on this class. The attribute called the_geom contains the geometry of each observed data.

RemoteObservation table contains the raster data of each observation that is contained in the Observation table, where observationType is remote. It has many-to-one relationship with the Observation table. Its relationship with the In-situObservation are executed on the fly through the geometry columns . Its attribute called rast contains the geometry or coordinate information as well as the the data values (geomval). The intersection between the in-situ and remote observations tables is made possible through the intersection of the 'the_geom' and the 'rast' which is done on the fly. Also more complex operations such as calculate, vectorise, intersect with vector can be performed on this class.

Metadata tables houses some important header data about any raster data contained in the RemoteObservations table. It has many-to-one relationship with the RemoteObservation table. It can be updated, selected from, listed etc. from the user end through an SQL- language based request. This table is created implicitly and encapsulated in the remote_observation table and is used to describe the coverages.

3.2. The ER-diagram and logical design of the database model

Figure 4 is the Entity Relationship diagram and logical design of the proposed heterogenous sensor database. The diagram shows the relationship logics between the tables for an effective physical implementation in PostGIS, leveraging the primary and foreign keys for seamless integration between the class. The relationship and integration of the In-situObservation and RemoteObservation tables are executed on the fly, leveraging their geometry columns and the coverage concept.

4. Integrating the Heterogeneous Sensor Database with the OGC Web Services

We propose an SQL-based Web Query Service, WQS that delivers SQL queries from the user end to the database in the web service . This service can be intergated and accessed from within the user's web or desktop application. This service provides the cleint the flexibility and ability to construct queries of extensive complexity which is delivered to the database for processing. In this case, aggregations, processing and analysis of remote and in-situ observations are carried out at the database backend. The result of the query can be delivered in different formats such as ASCII, GML, KML, TIFF, JPEG etc. in compliance with the OGC web mapping services, the WFS, WMS and WCS. The user specifies the formats of delivery on the query by using the PostGIS "ST_As*" function. ASCII or text results are delivered to the client directly from the database through the WQS. If the request result is to be delivered as a raster coverage, then the query result is a raster or a rasterised vector

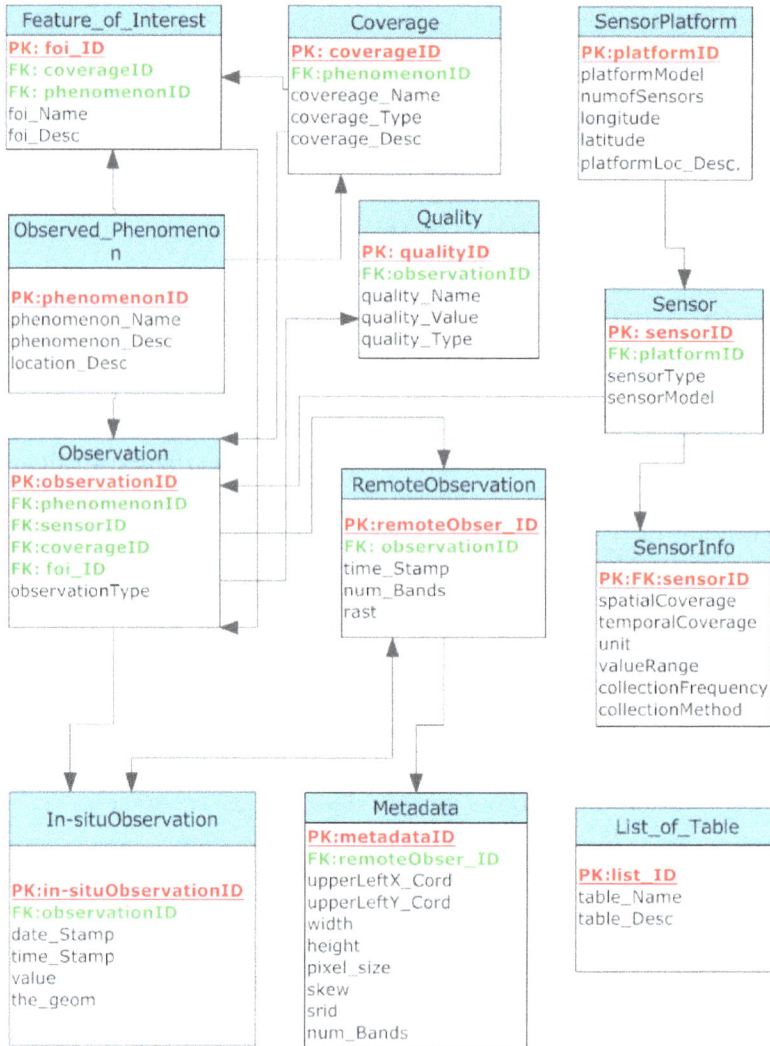

Figure 4: ER-Diagram and Logical Design of the Heterogeneous Database

and will be delivered to the client through the Web Coverage Service, WCS protocol. Similar process goes for a vector or vectorised query result which is delivered through the Web Feature service WFS protocol. The request result can be delivered as a JPEG or PNG image format to the user through the Web Map Service WMS protocol as described in figure 6.

4.1. The concept of the Web Query Service WQS

The Web Query Service, WQS is the proposed SQL query service that serves query from the client's web or desktop application to the heterogeneous sensor database. The Web Query Service delivers SQL queries from the client application through the network to the sensor database. It makes it easier to build and execute queries on a remote sensor database from any client application.

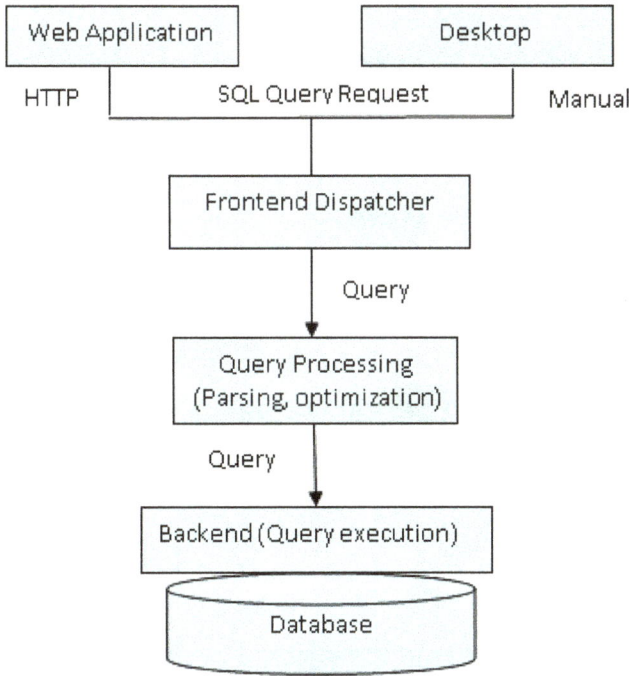

Figure 5: Conceptual Model of the proposed Web Query Service WQS

In Figure 5 the SQL query is delivered from the frontend dispatcher of client web or desktop application to the query processing and optimization module for optimization and parsing to the backend for query execution. From a web application, the SQL query request is dispatched via the HTTP. From within a desktop application, a connection to the database would have to be established manually before queries are sent to the database for execution.

Proposed Conceptual Architecture of Integrating the Heterogenous Sensor Database and OGC Web Services.

Figure 6 describes the conceptual achitecture of our proposed integration of the heterogenous sensor database as part of the Sensor Observation Service with the proposed Web Query Service WQS and other Web Services to deliver effective results to the end user.

The user on the client end, web or desktop application delivers SQL queries of any complexity through the WQS to the database. The result of the query is delivered back to the user through the relevant services depending on the format the result is requested. The ST_As * PostGIS function is used in the query to specify the format of delivery. When the user specifies for example ST_As GeoTIFF, the raster coverage query result is wrapped in XML and delivered to the client through the WCS protocol. The same process goes for query results specified in ST_As JPEG, PNG and KML or GML which are dilivered through the WMS and WFS respectively to the client. If no delivery format is specified in the query, the result is returned back to the client via the WQS by defualt in ASCII format. OGC web service operations such as GetCapabilities, DescribeSensor, DescribePlatform, GetObservation, DescribeCoverage or GetRaterMetadata, GetCoverage, ProcessCoverage etc. are carried out through this Web Query Service WQS by SQL queries.

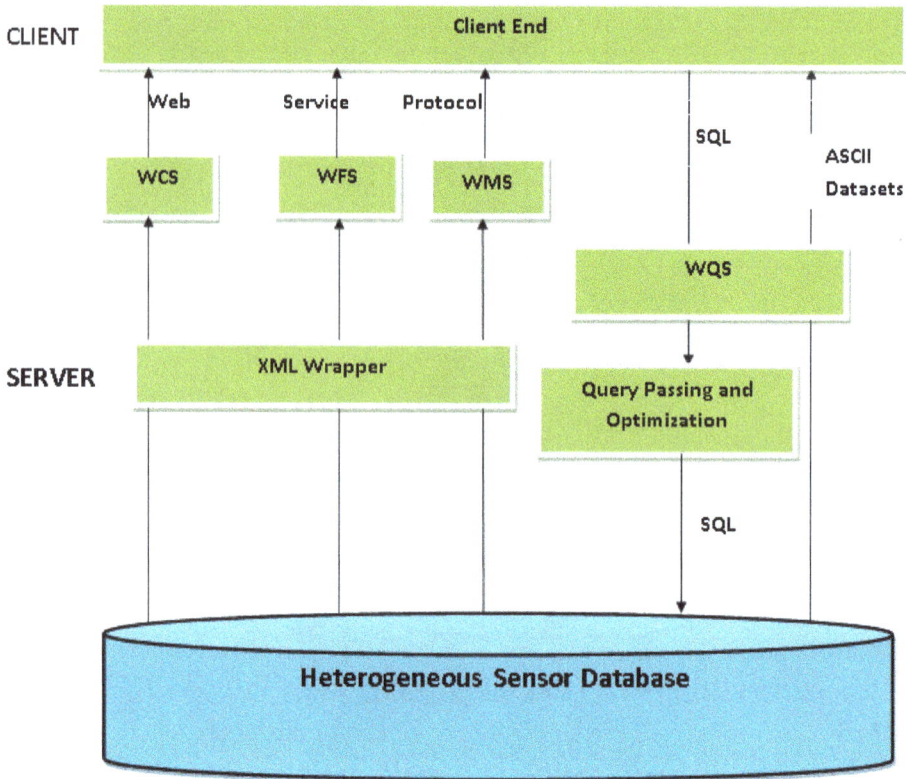

Figure 6: Proposed Conceptual Architecture of Integrating the Heterogeneous Database and the Web Services

5. Prototypical Implementation and Scenario Evaluation

In this section we did a prototypical implementation of the heterogeneous database model in PostGIS 2.0 as shown in figure 7. We loaded the tables with in-situ and remote sensor data as described in the logical model. In-situ sensor observations stored as vector coverages and remote sensor observations as raster coverages. In the heterogeneous database we had in-situ and remote sensor Land Surface Temperature LST coverage, Sea Surface Temperature SST coverage, Reference Evapotranspiration in-situ coverage, Normalised Difference Vegetation Index NDVI coverage and so on. Afterwards some few scenarios or use cases out of the numerous use cases where the proposed heterogeneous sensor database model can be leveraged to accomplish geo-scientific queries and processing involving remote and in-situ observations were carried out. The query scenarios were executed from a client desktop application (the OpenJUMP Desktop GIS application) after establishing a connection to the heterogeneous sensor database at the server. Scenarios ranging from a simple case where a geo-scientist would want to obtain the temperature difference between in-situ and remote temperature observations to a more complex case of estimating daily plant Evapotranspiration of a particular location.

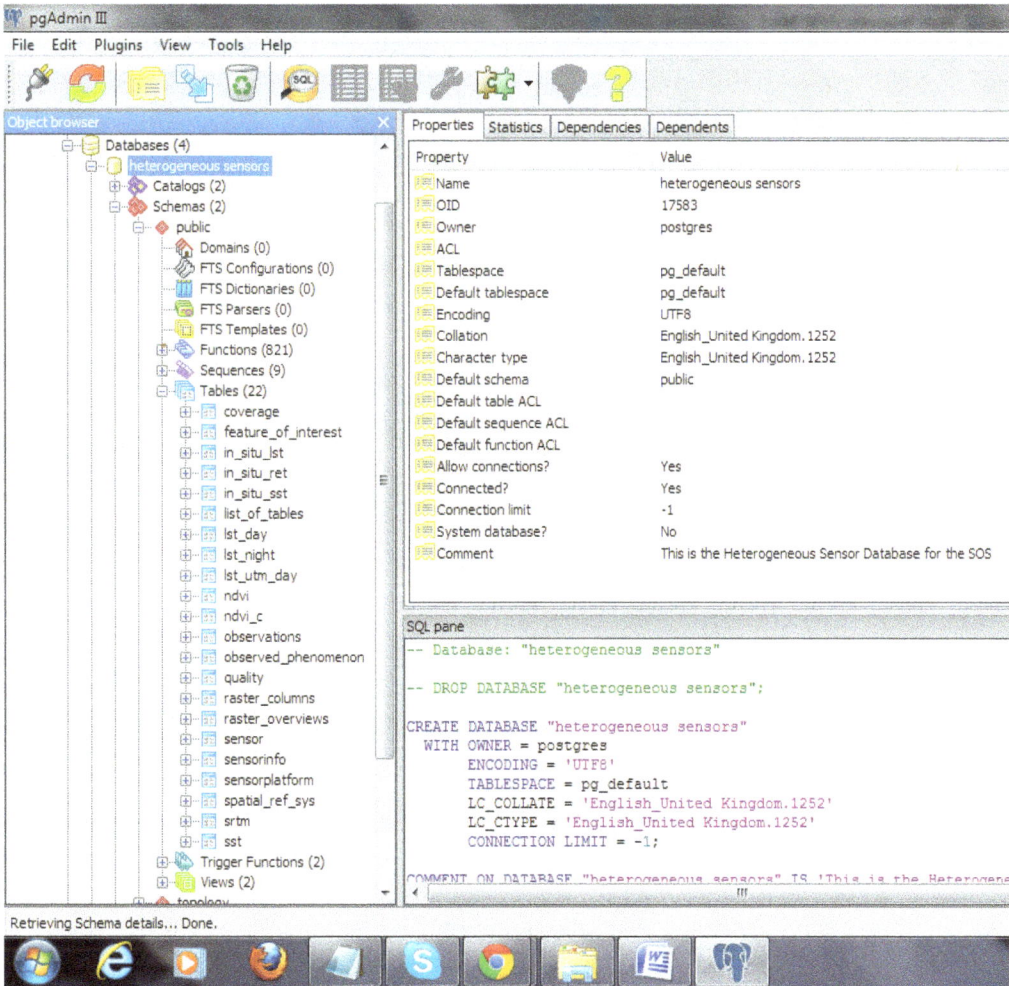

Figure 7: A screen shot excerpt of the heterogeneous sensor database model with the tables

Figure 7 is a screen shot excerpt showing the physical implementation of the database model in PostgreSQL/PostGIS 2.0 database management system. Both the remote and in-situ sensor observations efficiently stored for seamless integration.

5.1. Scenario 1: In-situ and satellite surface temperature analysis

This scenario calculates the temperature difference between the in-situ sensor land surface temperature observation and remote sensor land surface temperature observation of a particular location. Listing 8 was used to obtain the required result from within the OpenJump desktop application.

Listing 1: Scenario 1 implementation SQL code

```
SELECT val1, (gv).val AS val2 ,val1-(gv).val AS diffval,geom
FROM ( SELECT ST_intersection(rast,the_geom) AS gv,
temp_value AS val1, ST_AsBinary(the_geom) AS geom
FROM in_situ_lst , lst_day
```

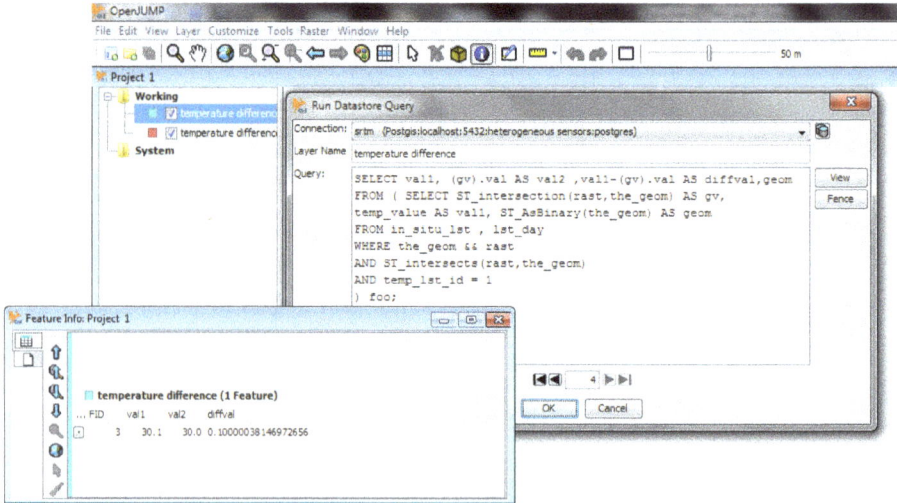

Figure 8: Screen short of Scenario 1 implementation in OpenJump

```
WHERE the_geom & rast
AND ST_intersects(rast,the_geom)
AND temp_lst_id = 1
) foo;
```

Here, this query picks up a particular temperature observation from the in-situ land surface temperature 'val1', in-situ_lst table of a location where id = 1, compares the temperature value with the corresponding remotely observed temperature, 'val2' of that same location on the raster temperature coverage, Lst_day and returns the difference, 'diffval'.

Figure 8 below is the implementation screen short excerpt from the OpenJump desktop application showing the connection to the heterogeneous sensor database and the result of the query from within the OpenJUMP client desktop application.

In figure 8 below, connection to the heterogeneous sensor database and the SQL query are depicted on the upper right hand side of the image while the query result on the lower left corner.

5.2. Scenario2: Estimation of Actual Crop Evapotranspiration ET at the Database Backend

We have the in-situ Reference Evapotranspiration RET coverage from weather automatic stations and NDVI raster coverage of the spatio-temporal attribute in the heterogeneous sensor database. Therefore we can calculate the Actual Evapotranspiration AET, from an aggregation of RET and Fraction of Vegetation cover FVC, where FVC is derived from NDVI [1].

```
AET = FVC * RE, [1]
FVC = N^2
N   = (NDVIp-NDVImin)/(NDVImax-NDVImin)
```

Where

```
AET       = Actual Evapotranspiration
FVC       = Fraction Vegetation Cover
RET       = Reference Evapotranspiration obtained from in-situ observation
NDVIp     = the NDVI Value at a point p
NDVImax   = the maximum NDVI value within the entire area of observation
NDVImin   = the minimum  NDVI value within the entire area of observation
```

In the query below, a geo-scientist can leverage the simple formula above to obtain the AET of a particular location, having the RET of that particular point on the in-situ observation table and the NDVI coverage of the area as well in the heterogeneous sensor database.

To implement this scenario, we could use 1 and 0 as the approximate maximum and minimum NDVI values respectively within the area, this would give us an approximate estimation not very precise. But to obtain the actual NDVImax and NDVImin of the coverage area, we used the SQL query below in listings 2 and 3, which can then be factorized in the comprehensive AET query statement in listings 4 to obtain the precise AET.

Listing 2: SQL Query to obtain the NDVImax

```
SELECT (stats).max}
  FROM (SELECT ST_SummaryStats(rast) AS stats
  FROM ndvi
 ORDER BY stats DESC
 LIMIT 1 ) AS foo;
```

Listing 3: SQL Query to obtain the NDVImin

```
SELECT (stats).min
  FROM (SELECT  ST_SummaryStats(rast) As stats
  FROM ndvi
 ORDER BY stats ASC
 LIMIT 1 ) AS foo;
```

In this our example case, we calculated the AET of a point in the RET, in_situ_ret table where id =1 by implementing the SQL statement in listing 14 below. From the query in listing 21 and 22, we obtained the NDVImax and NDVImin as 0.86 and 0 respectively and factorized them in as shown in the listing 4 below and got the results from within the OpenJUMP client desktop application shown in figure 9.

Listing 4: Scenario 4 implementation SQL code

```
SELECT RET, NDVIp,(pow(((NDVIp-0.86)/(0.86-0)),2)) AS FVC,
       (pow(((NDVIp-0.86)/(0.86-0)),2))*RET as AET, the_geom
  FROM (SELECT ST_Value(R.rast,I.the_geom) as NDVIp, I.value as RET,
       ST_AsBinary(I.the_geom) as the_geom
  FROM in_situ_ret I, ndvi R
 WHERE ret_id = 1
   AND ST_Value(R.rast,I.the_geom) IS NOT NULL) foo;
```

Lots of other scenarios were also tested for example, calculation of Weighted Mean surface temperature values from a vector buffer. A scenario where, one could select a particular

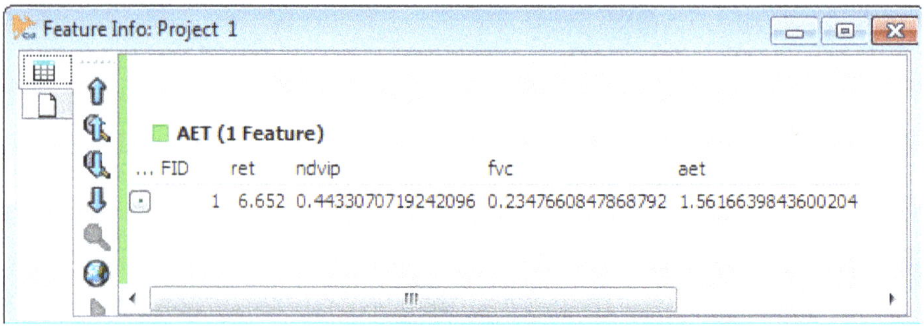

Figure 9: Screen short excerpt of a sample Scenario 4 implementation result in OpenJump client end

observation in the in-situ temperature observation, create a buffer of a given radius around that observation, then overlap this buffer geometry on the raster temperature coverage and obtains a weighted mean surface temperature value within the buffered region from the raster coverage. Also a scenario to describe a raster coverage metadata such as done in the OGC Sensor Web "DescribeCoverage" to obtain the metadata of a particular coverage. . In this case, we could leverage the PostGIS raster metadata description function to provide the client side description of a raster coverage through an SQL query.

In general the results of the sample queries shown above for the mentioned scenarios are alphanumeric or CSV formatted. They are returned to the client directly from the database. Other results formats are also possible as described in section 4.1 above, depending on how the client wants the results delivered.

6. Evaluation and Conclusion

In our final evaluation of the methods discussed we focus on three major topics, query flexibility, reduction of communication load and work and massive data retrieval load.

6.1. Query Flexibility

The various geo-processing scenarios we implemented in the prototypical implementation exercise from within the OpenJump client side desktop application show that, this approach of delivering SQL based queries from the client end direct to the database backend makes it more flexible for the user on the client end to deliver geo-processing queries of extensive complexity involving in-situ and remote sensor observations. Language based query request such as the (SQL) has been considered advantageous especially by the database community because is very flexible, declarative, optimizable and more safe in evaluation [10]. This extensive support for different kinds of geo-processing and analysis involving in-situ and remote sensor observations through native SQL queries makes this approach advantageous to the current approach of having different geo-processing modules on the web service for specific purposes. In that case users are restricted only to the specific geo-processing capabilities the service offers.

6.2. Reduced communication load

The prototypical implementation our proposed heterogeneous sensor database model and the model of how it can be integrated seamlessly with the OGC geo-web services show that these disparate sensor observations can be integrated and managed in a single spatial database leveraging PostGIS 2.0 functionalities. Hence communication load to different databases are invariably reduced. The communication time lag incurred in the downloading of raster images from a flat file database via the ftp, obtaining in-situ observations from an in-situ sensor observation service and integrating the two on the service middleware level is invariably reduced greatly, adopting this heterogeneous database approach.

6.3. Work and Massive Data Retrieval Load

Also taking a look at the contents and the processes in dynamic systems such as in [11], [2], [12], [13] and in the OGC Web Processing Services, they provide clients access and results based on pre-programmed calculations and/or computation models that operate on the spatial data. To enable geospatial processing and operations of diverse kinds, from simple subtraction and addition of sensor observations (e.g. the difference between satellite observed temperature and in-situ observation of a location) to complicated ones such as climate change models, requires the development of a large variety of models on the service middleware. This is massive in work load and huge amount of programming on the service. Also the data required for these services are usually retrieved dynamically from different databases and services which most times entails massive data retrieval especially from the satellite data (raster) storage.

Contrarily, by the means of a heterogeneous sensor database model such as developed and implemented in this research leveraging the functionalities of PostGIS 2.0 database extension, geo-processing and analytics involving remote and in-situ sensor data are carried out at the database backend by native SQL request statements. Therefore the variety of geo-processing work load on the service middleware is reduced. The service middleware in our case is majorly for service delivery from the client to database and vice versa. Massive data retrieval before processing is completely avoided. Also massive programming involved in the development of different kinds of geo-processing models on the web service is reduced.

7. Further Work

The practical usefulness of this proposed approach will be very more appreciated and leveraged when we are done with the full implementation of the model, integrating the proposed sensor heterogeneous database and other geo-web services in the SOS. This is our next milestone, to fully integrate this heterogeneous sensor database framework and the WQS with other geo-web services for query result delivery to the clients in different formats as described in figure 6.

References

[1] *Groundwater and Vegetation Effects on Actual Evapotranspiration Along the Riparian Zone and of a Wetland in the Republican River Basin.* **Gregory Cutrell, M. Evren Soylu.** Nebraska-Lincoln : s.n., 2009.

[2] *Development of a Dynamic Web Mapping Service for Vegetation Productivity Using Earth Observation and in situ Sensors in a Sensor Web Based Approach.* **Kooistra, L., et al.** 4, 2009, Sensors , Vol. 9, pp. 2371-2388.

[3] **Hamre, Torill.** Integrating Remote Sensing, In Situ and Model Data in a Marine Information System (MIS). *Marine Information System (MIS), in Proc. Neste Generasjons GIS.* 1993, pp. 181-192.

[4] *Geographic information -- Schema for coverage geometry and functions.* **ISO.** 2009, TC 211 - Geographic information/Geomatics.

[5] **PostGISWiki.** PostGIS UsersWiki. [Online] [Cited: September 10, 2011.] `http://trac.osgeo.org/postgis/wiki/UsersWikiCoveragesAndPostgis`.

[6] *Management of multidimensional discrete data.* **Baumann, Peter.** Issue 4, October 1994, The VLDB Journal, Vol. Volume 3, pp. 401-44.

[7] *Store, manipulate and analyze raster data within the PostgreSQL/PostGIS spatial database.* **Racine, Pierre.** Denver : `http://2011.foss4g.org`, 2011. FOSS4G.

[8] **postgis.refractions.net.** PostGIS 1.5.3 Manual. *postgis.refractions.net Web site.* [Online] [Cited: August 20, 2011.] `http://www.postgis.org/docs/`

[9] **Pierre, Racine.** WKTRasterTutorial01. *PostGIS.* [Online] June 2010. [Cited: 11 12, 2011.] `http://trac.osgeo.org/postgis/wiki/WKTRasterTutorial01`

[10] *Designing a Geo-scientific Request Language - A Database Approach.* **Baumann, Peter.** s.l. : Springer-Verlag Berlin, Heidelberg, 2009. SSDBM 2009 Proceedings of the 21st International Conference on Scientific and Statistical Database Management. ISBN: 978-3-642-02278-4.

[11] **Rueda, Carlos and Gertz, Michael.** Real-Time Integration of Geospatial Raster and Point Data Streams. *Statistical and Scientific Database Management.* 2008, pp. 605--611.

[12] *WARMER in-situ and remote data integration.* **AlastairAllen, et al.** Southampton (UK) : s.n., 30th March 2009. National Oceanography Center.

[13] *An integrated Earth sensing sensorweb for improved crop and rangeland yield predictions.* **Teillet, P M, et al.** 2007, Canadian Journal of Remote Sensing, Vol. 33, pp. 88-98.

## 16

Quantum GIS plugin for Czech cadastral data

Anna Kratochvílová and Václav Petráš

Students of Geoinformatics Programme
Faculty of Civil Engineering
Czech Technical University in Prague

Abstract

This paper presents new Quantum GIS plugin for Czech cadastral data and its development. QGIS is a rapidly developing cross-platform desktop Geographic Information System (GIS) released under the GNU GPL. QGIS is written in C++, and uses the Qt library. The plugin is developed in C++, too. The new plugin can work with Czech cadastral data in the new Czech cadastral exchange data format called VFK (or NVF). Data are accessed through VFK driver of the OGR library. The plugin should facilitate the work with cadastral data by easy search and presenting well arranged information. Information is displayed in the way similar to web applications, thus the control is friendly and familiar for users. The plugin supports interaction with map using QGIS functionality and it is able to export various cadastral reports. This paper provides ideas which can be generalized to develop QGIS plugin dealing with specific data.

Keywords: VFK, NVF, cadastre, ČÚZK, GIS, QGIS, OGR, C++, plugin

1. Introduction

1.1. Czech cadastre

Modern Czech cadastre has a long history with roots in Austria-Hungary in the 19th century. Since 2001 the Czech Cadastral Office for Surveying, Mapping and Cadastre (ČÚZK) has provided cadastral data in electronic form via the Information System of the Cadastre of Real Estate (ISKN) [1]. Many organizations and companies, both state and commercial, use this opportunity at an increasing rate. Several data formats exist, some of them comply with INSPIRE specifications [2].

The most used format is the Czech cadastral exchange data format called VFK (or NVF) which contains all data related to real estate. Unlike the other formats VFK contains information about ownership. Cadastral data in VFK are widely used by municipalities and state institutions for execution of their duties.

For using VFK data proper software is needed. Beside some proprietary programs, there have been attempts to provide free software solution: module v.in.vfk for GRASS GIS, Otevřený katastr (Open Cadastre) [3] for import to PostGIS and VFK driver [4] in OGR library. The last-named has the advantage that GDAL/OGR library is used by many programs, both proprietary and free open source. However, typical end-user is not able to access the functionality without some wrapper application providing graphical user interface (GUI).

For working with cadastral data which are basically spatial data an ideal environment is a Geographical Information System (GIS).

1.2. QGIS

For the development of the application we decided to use Quantum GIS (QGIS) [5]. QGIS has many advantages both for the user (e.g. customizable and easy to use GUI) and the developer (e.g. good API, simple plugin system).

QGIS is a free open source GIS licensed under GNU GPL[1]. QGIS is written in C++ programming language and uses Qt framework[2]. There are several possibilities how to add new functionality to QGIS. Firstly, a new plugin for QGIS desktop application can be written in C++ or Python. Secondly, it is possible to build your own application based on QGIS library. Finally, you can directly modify existing QGIS application.

The presented plugin was developed for QGIS Desktop, however also other components of QGIS exists (e.g. QGIS Server, QGIS Browser). In order to develop a plugin for QGIS it is necessary to use QGIS API (Application Programming Interface). The API is well documented and a guide for plugin development is available, too.

Thanks to GPL license everyone who obtains QGIS application can obtain also its source codes. As a result one can prove how performed analyses are implemented which is crucial for academic work [6]. For public administration, low cost of free open source software can be a decision criterion. Efforts to introduce free open source solutions to public administration can be found on European level.[3]

2. VFK data format and OGR-VFK driver

2.1. VFK data format

The Czech cadastral exchange data format called VFK (or NVF) replaced previous format VKM in 1996[4]. Files in this format are provided by the Czech Office for Surveying, Mapping and Cadastre. VFK contains information about real properties (parcels, buildings, building units), including their description, geometry and related property rights. In contrast to other current Czech cadastral formats (e.g. INSPIRE Cadastral Parcels), VFK contains information about owners. VFK files are provided to public for fee, however municipalities and state institutions obtain it free of charge [7].

From the technical point of view VFK is a text format which resembles CSV format (see below).

```
&HPAR;ID N30;STAV_DAT N2;DATUM_VZNIKU D;...
&DPAR;1319150210;0;"23.06.2003 14:54:44";...
&DPAR;1319151210;0;"24.04.2002 09:26:01";...
```

---

[1]GNU General Public License, http://www.gnu.org/licenses/licenses.html
[2]http://qt.nokia.com
[3]https://joinup.ec.europa.eu/
[4]http://www.cuzk.cz/Dokument.aspx?PRARESKOD=998&MENUID=0&AKCE=DOC:10-SDELENI_K_SVF

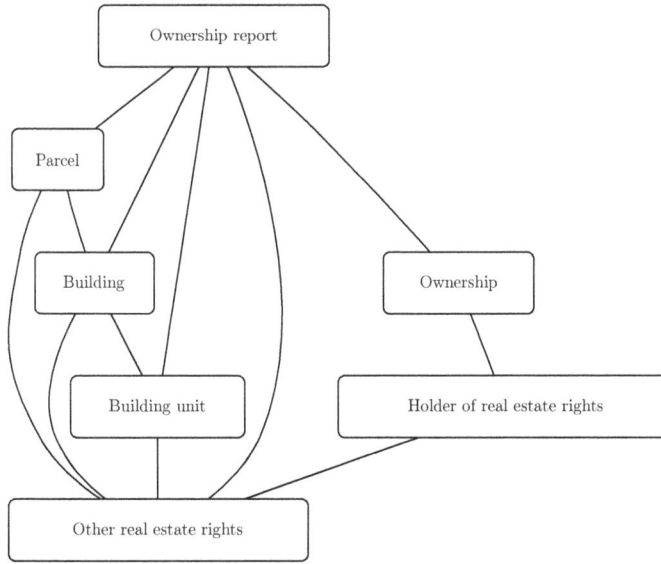

Figure 1: Simplified relations between entities in database (for full schema please refer to [9])

Official format description and underlying database schema is provided, however this documentation [8] is not sufficient for building queries. The description provided in [9] was used as a reference for the plugin development. For schema overview refer to figure 1.

2.2. OGR-VFK driver

OGR[5] is an free open source C++ library which enables read (and in certain cases also write) access to various geospatial vector formats including ESRI Shapefile, PostGIS or Oracle Spatial. OGR is the part of GDAL library, therefore it is also referred as GDAL/OGR. This library is used in many free open source software projects like GRASS GIS, QGIS or MapServer and in proprietary software (e.g. ESRI ArcGIS) which is possible thanks to MIT style free software license. The support of the Czech cadastral exchange format (VFK) was missing till 2010 when M. Landa [4] implemented the VFK driver which has then become the part of the library. Thanks to the driver every software using OGR library can access Czech cadastral data in VFK format.

During the QGIS VFK plugin development the VFK driver was improved by its author in order to reflect needs of the plugin. The most significant enhancement is the export to SQLite3 file database. This file is then used by the plugin but it can be accessed by any SQLite browsing and editing tools.

3. Implementation

3.1. QGIS plugin for Czech cadastral data

QGIS plugin for Czech cadastral data is primary intended for using by local municipalities. The implementation of VFK driver in OGR library enabled to view Czech cadastral data in

---

[5]http://www.gdal.org/ogr/

QGIS, however it was still inconvenient to browse the non-spatial but significant part of the data which is represented by a large and complex set of attribute tables. The aim of the plugin is to facilitate the work with this kind of data which can be then viewed and analysed in relation to the spatial part.

The development of the plugin was split into two parts. During the first part we developed an application not connected to QGIS. This involved the development of the crucial functionality — database search in SQLite database created by VFK-OGR driver, and generating and exporting various reports. During the second part the application was connected to QGIS as a plugin so that the map interaction could be implemented.

Data are handled by the plugin in the following way. QGIS reads spatial data from VFK file through VFK-OGR driver. The driver creates SQLite file, from which plugin reads attributes of spatial features and other related non-spatial data. Plugin connects these two sets of data through unique identifiers defined in VFK documentation [8].

In order to use the plugin functionality user is supposed to open the VFK file from within the plugin instead of using the QGIS standard dialog for adding vector layers. This enables to optimize loading process comparing to standard QGIS OGR layer loading process by avoiding calling certain unnecessary procedures.

Another improvement affecting the performance is the fact that the driver uses the SQLite database file from previous run if available. During the plugin development crucial attribute columns were identified and VFK-OGR driver now creates database indices for this columns. Indices improve the speed of attribute querying in the plugin. These changes were implemented by the author of the VFK-OGR driver.

The plugin was developed with development version of QGIS and development version of OGR.

3.2. Plugin functionality

Plugin is in the stage of first prototype and its functionality is so far limited. However, it contains basic functionality to solve all common tasks. This functionality includes search according to various parameters depending on searched feature or object. Currently it is possible to search information about parcels, buildings and owners. Several reports are available:

- report about parcels
- report about buildings
- report about building units (flats or non-residential space)
- report about owners
- ownership reports (according to Czech cadastre style)

These reports are interconnected (figure 2) so that the user can easily get from one report to another.

Plugin contains a browser similar to the standard web browser. This browser enables interactive browsing of attribute data in the following way:

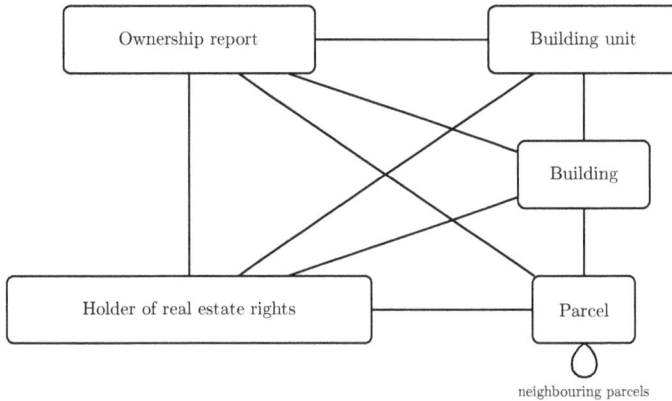

Figure 2: Interconnections between the reports

- Data are represented as an styled HTML page.

- Hyperlinks are used for navigation through various reports.

- Navigation includes buttons Back and Forward.

- Browsing history of visited pages does not require to redo database queries.

Plugin provides the possibility to show current state of parcels and buildings in web application Viewing Cadastre (Nahlížení do katastru nemovitostí, ČÚZK application) which provides limited access to cadastral data with map interface. This application is launched in system web browser showing currently selected feature. Unlike Viewing Cadastre, VFK plugin provides possibility to search by owner (e.g. find all real estates of one owner).

Reports generated by the plugin can be exported into two formats — HTML with CSS stylesheet and LaTeX(enables creating PDF). HTML can be easily imported to OpenOffice.org or LibreOffice so that the document structure is preserved.

In order to use information from database search or search information about features selected in map the following functionality for synchronizing was developed:

- Parcels and buildings which are currently shown in plugin browser can be selected (highlighted) in the map.

- Information about selected feature(s) in the map can be shown in plugin browser.

For convenience, cadastral map layers are displayed with predefined style. Most importantly it includes displaying parcel numbers with special cadastral formatting (style originally coming from Austria-Hungary).

Brief help page is embedded in the plugin. It contains hyperlinks to find wanted functionality easily.

It is worth noting that when using the plugin user can profit from all the functionality provided by QGIS too (e.g. loading WMS layers).

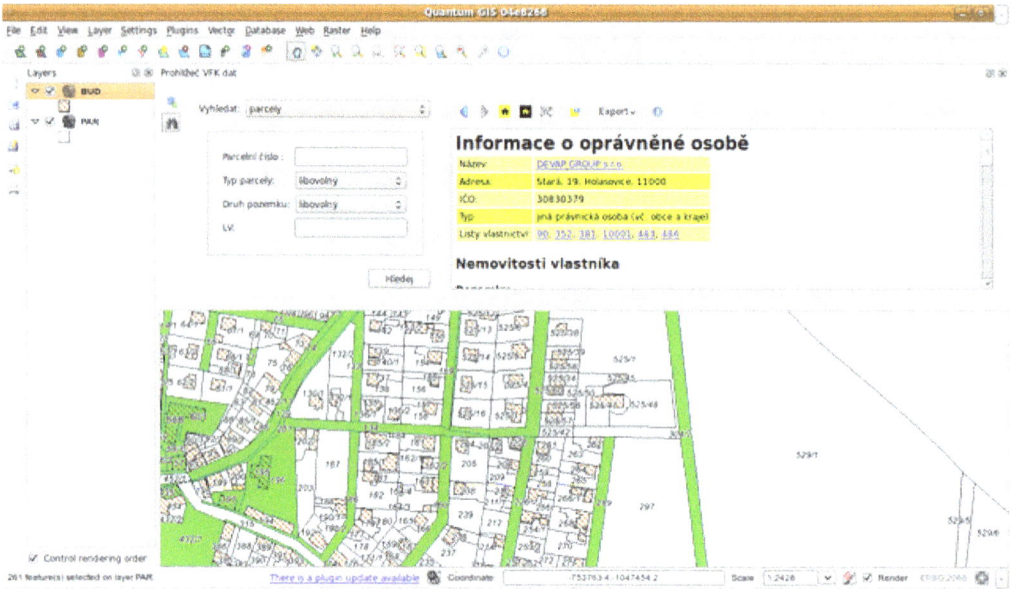

Figure 3: Plugin showing information about parcels' owner

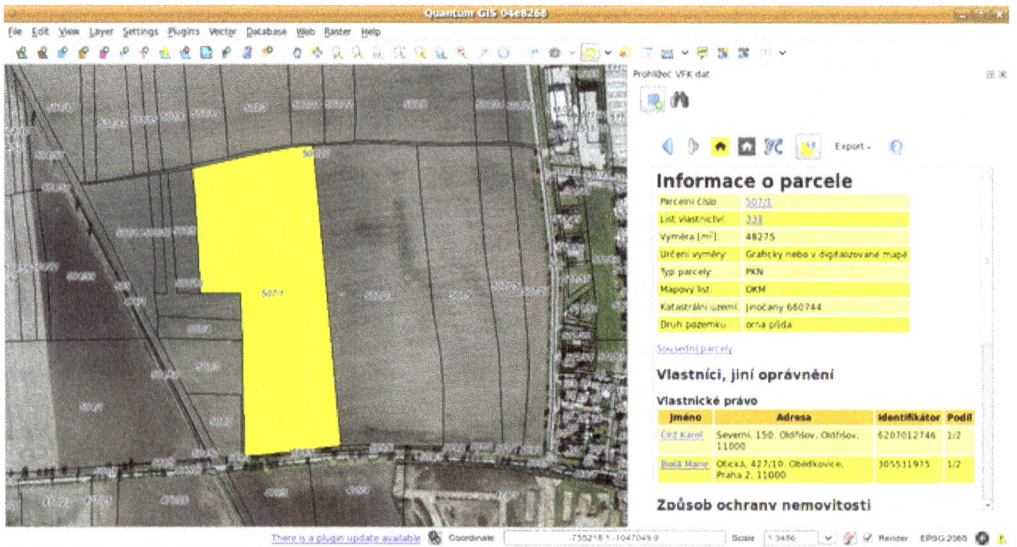

Figure 4: Docked plugin window with hidden control panel

3.3. Plugin GUI

The graphical user interface (GUI) of the plugin is divided into two main parts — the browser showing data reports and the control panel with toolbar for data import and search. The GUI was designed so that new functionality could be added easily without making the GUI cumbersome.

Thanks to powerful Qt framework, the plugin window can be floating or docked anywhere in QGIS application window (see figures 3 and 4). This is particularly advantageous to users using large screens. This way user can see both the map and related data.

4. Further development

Further development will be based on results of testing. Nevertheless, there are some improvements which are already planned:

- the direct support of other output formats (PDF, ODF)

- export of geometry

- using threads for time consuming data loading and querying (make the user interface responsive)

- database file handling

- add more layer styles

5. Conclusion

The Czech Cadastral Office for Surveying, Mapping and Cadastre manages large database containing both spatial and non-spatial data. These data have been used increasingly since launching Internet based Remote Access in 2001 [1]. The majority of clients is coming from public administration which gets this data free of charge [7].

Presented plugin aims to local municipalities which need tools for accessing these data. They can gain advantage from solution which uses free open source software. Apart from money saving, they avoid dependency on one certain software supplier and they can get software which respects their needs [10]. The plugin makes use of two free open source projects QGIS and GDAL/OGR to provide functionality needed for handling Czech cadastral data. It uses GIS environment to interactively browse spatial and attribute data and bind them together to get clear overview.

6. Acknowledgement

We are grateful to the author of VFK-OGR driver, Ing. Martin Landa, for additional enhancements of the driver needed for the plugin. We also thank to Mr. Jiri Sobotik from municipality Novy Jicin who encourages us to start the plugin development and contributes with his ideas about plugin functionality and testing.

7. References

1. ČESKÝ ÚŘAD ZEMĚMĚŘICKÝ A KATASTRÁLNÍ. Annual Report 2011. ČÚZK, 2012. ISBN 978-80-86918-66-2. URL: http://www.cuzk.cz/GenerujSoubor.ashx? NAZEV=10-EVZ2011

2. SOUČEK, Petr and Jiří FORMÁNEK. Data spravovaná resortem ČÚZK jsou stále přístupnější. In: GIS Ostrava 2012 - Současné výzvy geoinformatiky. URL: http: //gis.vsb.cz/GIS_Ostrava/GIS_Ova_2012/sbornik/papers/soucek.pdf

3. JEDLIČKA, Karel, Jan JEŽEK and Jiří PETRÁK. Otevřený katastr – svobodné inter-netové řešení pro prohlížení dat výměnného formátu katastru nemovitostí. In Geoinfor-matics FCE CTU. Praha: ČVUT, 2007. p. 111-117. URL: http://geoinformatics. fsv.cvut.cz/gwiki/Geoinformatics_FCE_CTU_2007

4. LANDA, Martin. *OGR VFK Driver Implementation Issues.* In: Proceedings – Sympo-sium GIS Ostrava 2010. p. 8. ISBN 978-80-248-2171-9, ISSN 1213-239X. URL: http: //gis.vsb.cz/GIS_Ostrava/GIS_Ova_2010/sbornik/Lists/Papers/EN_1_10.pdf

5. Quantum GIS Development Team, 2012. *Quantum GIS Geographic Information Sys-tem.* Open Source Geospatial Foundation Project. URL: http://qgis.osgeo.org

6. ROCCHINI, Duccio and Markus NETELER. Let the four freedoms paradigm apply to ecology. *Trends in Ecology.* 2012, 27, vol. 6, p. 310-311. ISSN 01695347. DOI: 10.1016/j.tree.2012.03.009. URL: http://www.sciencedirect.com/science/article/ pii/S0169534712000742

7. Česká republika. Zákon České národní rady ze dne 7. května 1992 o katastru nemovi-tostí České republiky (katastrální zákon). In: *Sbírka zákonů České republiky.* 1992. URL: http://portal.gov.cz/zakon/344/1992

8. ČESKÝ ÚŘAD ZEMĚMĚŘICKÝ A KATASTRÁLNÍ. *Struktura výměnného formátu informačního systému katastru nemovitostí České republiky* [online]. 23. 2. 2012 [cit. 2012-04-07]. URL: http://www.cuzk.cz/GenerujSoubor.ashx?NAZEV=10-D12U.

9. LANDA, Martin. *Návrh modulu GRASSu pro import dat ve výměnném formátu ISKN.* Master thesis. 2005. ČVUT Praha. URL: http://gama.fsv.cvut.cz/~landa/publications/ 2005/diploma_thesis/martin.landa-thesis.pdf.

10. FOGEL, Karl. *Tvorba open source softwaru: Jak řídit úspěšný projekt svobodného soft-waru.* 2010. CZ.NIC, 2012. ISBN: 978-80-904248-5-2. URL: http://knihy.nic.cz/ files/nic/edice/karl_fogel_poss.pdf.

## 17

Minimal Detectable Displacement Achievable by GPS-RTK in CZEPOS Network

Martin Raška[a] and Jiří Pospíšil[b]

[a]Katastrální úřad pro Karlovarský kraj, Katastrální pracoviště Sokolov
Boženy Němcové 1932, 356 01 Sokolov, Czech Republic
`martin.raska@cuzk.cz`

[b]Faculty of Civil Engineering, Department of Special Geodesy
Thákurova 7, Praha 6 16629, Czech Republic
`pospisil@fsv.cvut.cz`

Abstract

In this paper we have made a brief study of RTK precision to estimate possibilities of network RTK using CZEPOS for purposes of geotechnic monitoring of landslides in real time. In this paper we describe a testing methodology, which resulted in estimation of point-position precision and describing minimal detectable positional change. Based on our results it is concluded that displacements could be detected with centimetre accuracy even with short-period observations.

Keywords: GNSS, network RTK, landslide monitoring

1. Introduction

The measurements with GNSS are easier than terrestrial methods for displacement monitoring of points [12]. As GNSS technology and short observation time, followed up by terrestrial observation, usually provides sufficient precision for rough position precision requirements, as shown in experimental geomorphologic analysis [9], with increasing demands on precision, usually terrestrial methods are preferred, despite their complicated and time consuming procedure [1]. Unlike to RTK, the position of observed points with permanent stations may by determined with millimetre accuracy [6].

In 2007, 27 permanent GPS reference stations comprised the CZEPOS reference system. They cover almost the whole area of the Czech Republic, however some border areas extend outside this network (see Fig. 1), 23 of them owned and maintained by the Land Survey Office, whilst the others are supported and maintained by research facilities. At the end of 2008, the whole network had been connected with surrounding networks in Germany, Austria, Poland and Slovakia, thus computing corrections in border areas of the Czech Republic has become more reliable. In December 2009, reference station in Moravský Krumlov has been relocated to Znojmo. Besides storing continuous data for post-processing computing (available through the web pages of the Land Survey Office [5]), real-time data are provided as well. Data flow could be divided into two parts - DGPS corrections (code only, for GIS applications, decimetre precision) and RTK phase based corrections (centimetre precision). In this paper only the RTK service will be discussed.

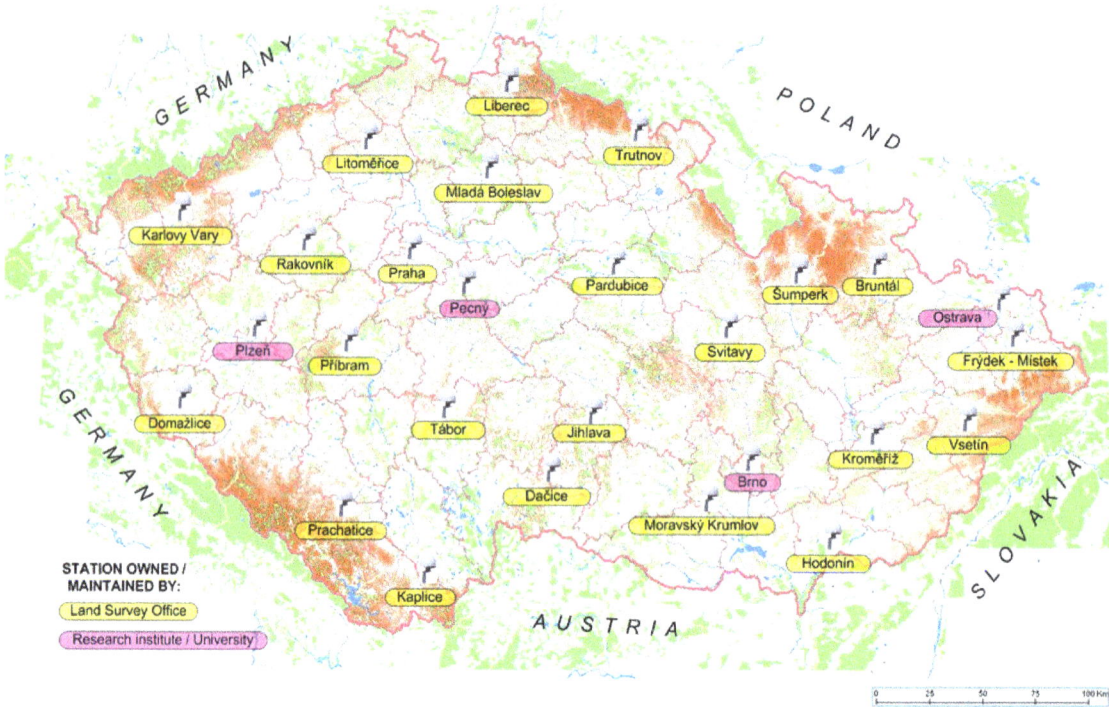

Figure 1: Reference stations of CZEPOS network

Basically, three different kinds of position solution are provided: Simple RTK, RTK-FKP and RTK-PRS. In "Simple RTK" the user chooses one of the reference stations, and corrections are then computed from that reference station only. This method is the least precise, as accuracy of baseline vector determination decreases with increasing length, as shown in [2]. Method is used only with very short baseline vector lengths or under some special circumstances (the next two solutions are computed as network solutions thus during a network solution failure, only "simple" corrections from individual reference stations are available). In "RTK – FKP", the nearest reference station is chosen automatically, but corrections are computed from modelling of the whole network area (data from all CZEPOS stations are processed). The last service provided is "RTK – PRS". In fact, PRS (pseudo-reference station) is the other name for virtual reference station (VRS), where the user gets corrections and/or observables from a virtual reference station [2], based on his initial position (NMEA) provided by a GNSS rover. Observables are computed for the reference station virtually located approximately 5 km from rover position, in the direction to the nearest reference station. Observables are created from network model of corrections, taking all observations in the whole network into account. Several experiments [10] proved that the precision of a VRS position is similar (thus being in order of 10^{-2} m) to those achieved by Simple RTK.

Corrections are received by a GPRS modem, usually integrated in a GNSS receiver, but today only the GPRS protocol is being used, while the whole area of the Czech Republic is covered by GSM signals and the GPRS data rate (theoretically with maximum at hundreds of kbps, [13]) meets the requirements for on-line data processing.

As common in such applications, the international standard Radio Technical Commission for

Maritime Services (RTCM) version 2.3 is currently used. In February 2009 the Land Survey Office introduced improved version of RTCM (version 3.1), which supports GLONASS correction transfer and has a more effective structure. With the last version of RTCM message [15], a new service "VRS3-MAX" became operational. With this service, observables and corrections for a virtual reference station are computed from a few (usually six) reference stations surrounding the rover. The closest one (called "Master") provides correction data and the others (called "Auxiliary") provide correction differences. There is an alternative service "VRS3–iMAX" for older types of receivers, where the user receives only correction data from the Master station, but Auxiliary stations data are already included in computation. Messages with corrections are sent via Networked Transport of RTCM via Internet Protocol (NTRIP) as needed for data transfer using Internet (as described in [15] and [14]).

2. Background theory

Coordinate system in the Czech Republic

For land surveying purposes, the Czech Republic uses its own national coordinate system (abr. S–JTSK). It could be described as two axes perpendicular to each other, axis +Y heading to west, axis +X heading to south. The origin is chosen so that the X–coordinate value is always greater than the Y–coordinate value, with horizontal angles measured in a clockwise direction. The height coordinate system (abr. Bpv), is a combination of levelling and gravity observations, which are used to derive heights (according to Molodensky's theory [7]).

The European continent drifts approximately 2-3 cm/year so the CZEPOS system primarily works with European Terrestrial Reference System (ETRS–89) instead of WGS84. Thus 3D–transformation from ETRS89 to S-JTSK (and Bpv) is required and is discussed below.

Datum transformation

Different solutions for transforming ETRS89 coordinates to the national system S–JTSK using universal transformation parameters already exist. The one used in Trimble GPS devices proved unusable in real applications due to distortion of the S–JTSK. This leads to very large residual vectors between "computed" (using global transformation) and "known" (computed from terrestrial observations) coordinates of points. Thus a local transformation is commonly being used. The experiment described below used a local transformation with 5 identical points (points with known coordinates in both the ETRS89 and S-JTSK system). In June 2008, Trimble introduced an improved global transformation algorithm, based on prof. Kostelecky's work [3]. Despite having a "random" characteristic of residual vector headings (compared with the original Trimble transformation), results (computed for the area surrounding city of Karlovy Vary) are better in scale by approximately one order (as we can see at Fig. 2).

The coordinates of all identical points used for a transformation, both ETRS89 and S–JTSK obtained from the Land Survey Office Database, have been used. As horizontal and vertical coordinates in national systems come from two different sources of observations and computations, local transformation from ETRS89 to S-JTSK (and Bpv) are strongly recommended to be separated into two standalone parts, horizontal and vertical (practical applications showed that using a geoid model for vertical transformation proved useful, while the geoid model

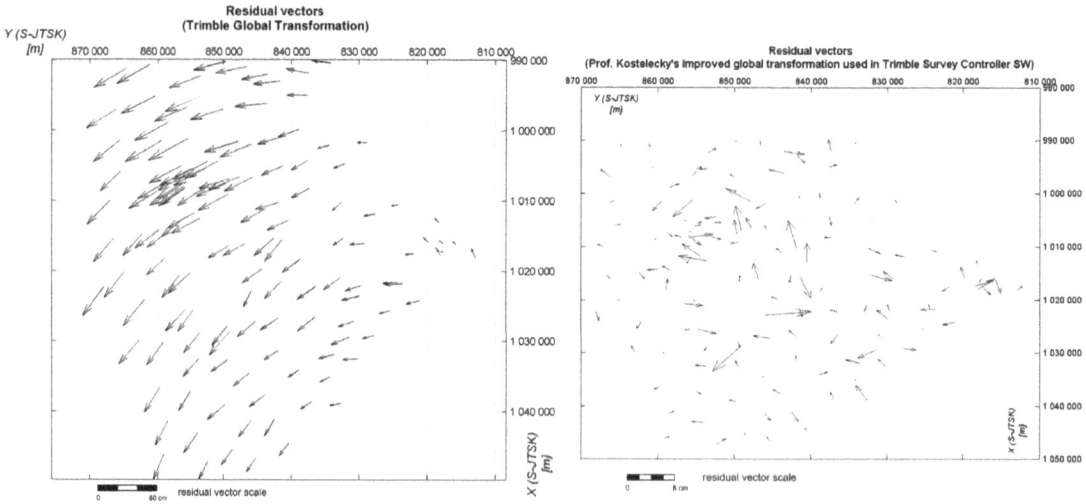

Figure 2: Comparing residual vectors of global transformation

for the Czech Republic [4] gives better results than generally used EGM96). Announced in October 2010, reference frame of the ETRS (ETRF89) is replaced by the ETRF2000.

3. Experiment overview

We chose three different sites A-C (with different terrain morphology), see Fig. 3 (represented by surveying landmark – granite stones with a diagonal cross on their top), located inside the area defined by identical-points to be used for the datum transformation. Each observation at one site included three kinds of CZEPOS data receiving (RTK only, RTK–FKP or RTK–PRS), each one for three different observation lengths (default options of receiver) periods (1 s, 5 s and approximately 180 s). For statistical analysis of the data we stored S-JTSK coordinates, as well as *a priori* accuracy characteristics. In the next step we used the entire data record for an empiric estimation of precision. During the experiment the only device used was a GPS rover Trimble R8.

Precision estimation

As the aim of this experiment was to describe the possibilities of RTK for point position change, there was no interest in deriving absolute positional accuracy in S–JTSK. Only the precision of repeatedly observed point has been tested.

As mentioned earlier, the whole data set was used for computation of positional precision estimation. During the testing procedure in three different areas (with different terrain morphology), possible systematic error (caused by specific environmental conditions) was significantly reduced. After obtaining a set of coordinates, adjustment using the least square method was used. In fact, it led to a computation of the weighted mean value for each coordinate (it was impossible to process all coordinates at one go due to their correlation; three coordinates – Y, X, H from each observation are correlated as being computed from the same set of pseudoranges).

We solved (or at least reduced) problems with different satellite constellations by introducing

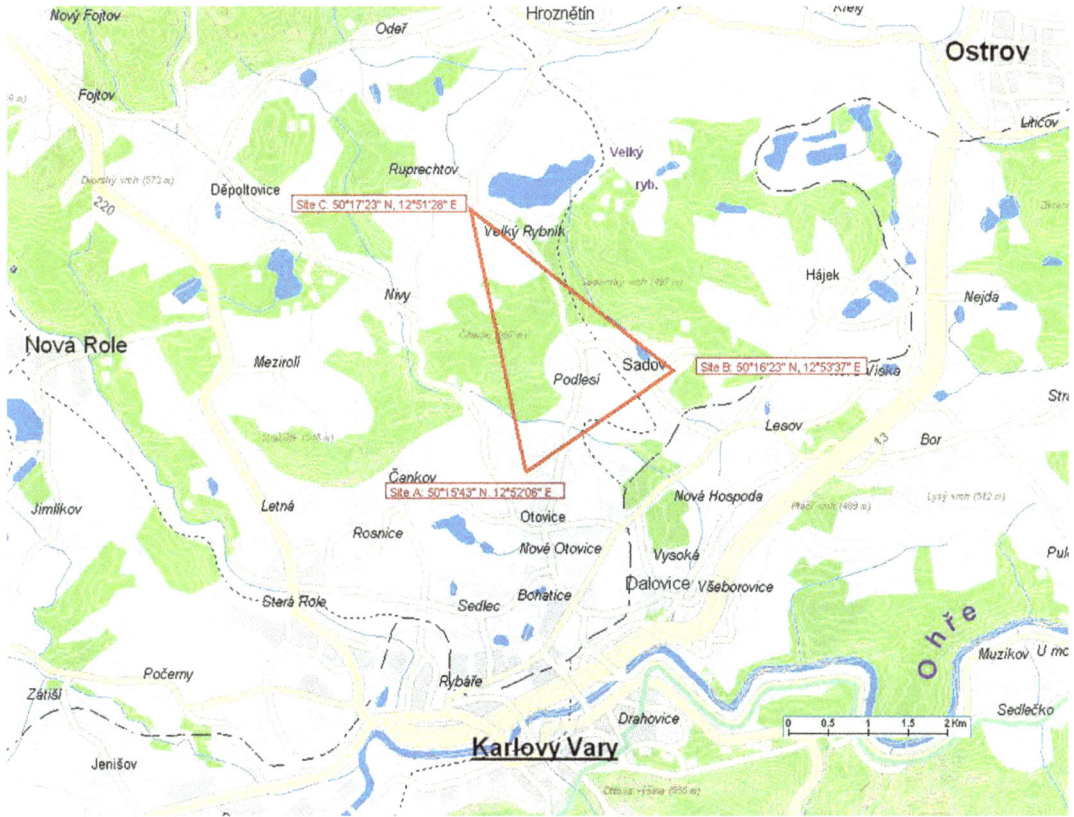

Figure 3: Geometric configuration of sites occupied during experiment

a Relative Dilution of Precision (RDOP) value for the deriving weights, as it takes into account satellite-receiver(s) geometric configuration as well as its changes during observation and observation interval [16]. Progress of RDOP value (based on model using testing site A, reference station in Karlovy Vary and 4 visible satellites) is shown at Fig. 4. It is clear, that unlike PDOP, RDOP value depends not only on instant satellite configuration, but on configuration-change during observation and observation length as well.

The next input parameter used to derive weights was the root mean square (RMS), computed during the observation. The RMS indicates the quality of the (ambiguity) solution based solely on the measurement noise of the satellite ranging observations. Using this value helped us to implement several random and systematic errors, such as multipath.

As we could derive horizontal and vertical interpretations of RDOP (called HDOP and VDOP [2]), weights for observation i can be expressed as:

$$p_{X,i} = p_{Y,i} = \frac{2}{RMS_i^2 \, HDOP_i^2} \tag{1}$$

$$p_{H,i} = \frac{1}{RMS_i^2 \, VDOP_i^2} \tag{2}$$

The next step has led to computation of residuals v, using a weighted average, and deriving a value of the "general" standard deviation (for n' as number of degrees of freedom):

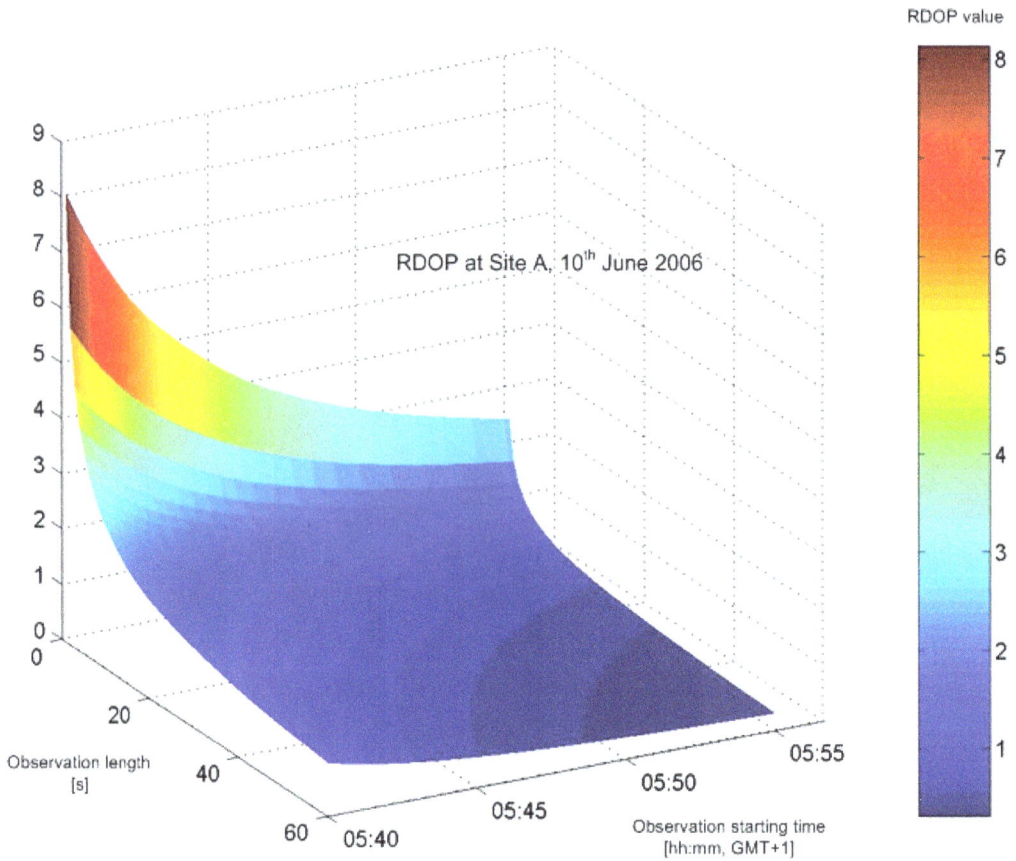

Figure 4: RDOP value modelling

The next step has led to computation of residuals v, using a weighted average, and deriving a value of the "general" standard deviation (for n' as number of degrees of freedom)

$$s_0 = \sqrt{\frac{\sum\limits_{i=1}^{n} p_i v_i^2}{n'}} \tag{3}$$

Thus the standard deviation (used for estimation of RTK precision) can be expressed as

$$s_X = s_{X_0}\sqrt{\frac{RMS_i^2\ HDOP_i^2}{2}} \tag{4}$$

$$s_Y = s_{Y_0}\sqrt{\frac{RMS_i^2\ HDOP_i^2}{2}} \tag{5}$$

$$s_H = s_{H_0}\ RMS_i\ VDOP_i \tag{6}$$

Three coordinates, X, Y and H, obtained in one observation epoch, are correlated. As the correlation relations between them are unknown, a precision estimation has been performed separately for each coordinate (Table 1).

Example (1): Considering values RMS=2,7 mm, HDOP=1.26, VDOP=1.93 (mean values of all RMS and DOPs achieved during experiment performance), experimental outcome yields

Table 1: Unit factors of coordinate standard deviations.

Correction Data source	Coefficient		
	s_{Y_0}	s_{X_0}	s_{H_0}
RTK-FKP	2.20	3.93	2.56
RTK-PRS	3.19	5.72	2.74
RTK	1.99	4.43	3.80

the resulting standard deviations for RTK-FKP service as $s_Y = 5.3$ mm, $s_X = 9.5$ mm and $s_H = 13.3$ mm.

Position change (accuracy estimation)

As we are able to estimate precision of single point coordinates, we could estimate accuracy of distance computed from 3D-coordinates of two points. Thus we could derive single point position change with its accuracy and estimate a minimum position change, which we are able to determine with a certain level of risk.

Using (4), (5), and (6) and the error propagation law, we can derive the standard deviation of such point position change

$$\Delta = \sqrt{(X_2 - X_1)^2 + (Y_2 - Y_1)^2 + (H_2 - H_1)^2} = \sqrt{\Delta_X^2 + \Delta_Y^2 + \Delta_H^2} \tag{7}$$

$$s_\Delta^2 = \frac{1}{2\Delta^2} \left[(RMS_2^2 \; HDOP_2^2 + RMS_1^2 \; HDOP_1^2)(s_{X_0}^2 \Delta_X^2 + s_{Y_0}^2 \Delta_Y^2) \right.$$
$$\left. + 2s_{H_0}^2 (RMS_2^2 \; VDOP_2^2 + RMS_1^2 \; VDOP_1^2)\Delta_H^2 \right] \tag{8}$$

Indexes 1 and 2 represent two different epochs of observation at one site. If we suggest a using significance level $\alpha = 5$, we could derive the minimum detectable point displacement as:

$$\Delta_{min} = 2 \, s_\Delta \tag{9}$$

As we could see, its value is closely related to HDOP/VDOP and RMS values during observation in two epochs. This value depends on the direction of position change.

Example (2): Considering values RMS=2.7 mm, $HDOP_1 = HDOP_2 = 1.26$, $VDOP_1 = VDOP_2 = 1.93$, Fig. 5 shows graph describing value Δ_{min} in relation to azimuth and elevation of point position change for service RTK-FKP.

If we plot the surface built up by minimal detectable displacement vectors, we obtain a 3D analogy of 2D Helmert's curve. To simplify (8), we could split the displacement vector into two parts, describing displacement in the horizontal plane and the vertical direction. By introducing a mean standard deviation of coordinates in horizontal plane

$$s_{X_0,Y_0}^2 = \frac{1}{2} \left(s_{X_0}^2 + s_{Y_0}^2 \right) \tag{10}$$

we are able to derive minimal detectable displacements in horizontal plane and vertical direction and have a better estimation of displacement precision in horizontal plane for different data sources of corrections from the CZEPOS

$$\Delta_{min,H_z} = s_{X_0,Y_0} \sqrt{RMS_1^2 \; HDOP_1^2 + RMS_2^2 \; HDOP_2^2} \tag{11}$$

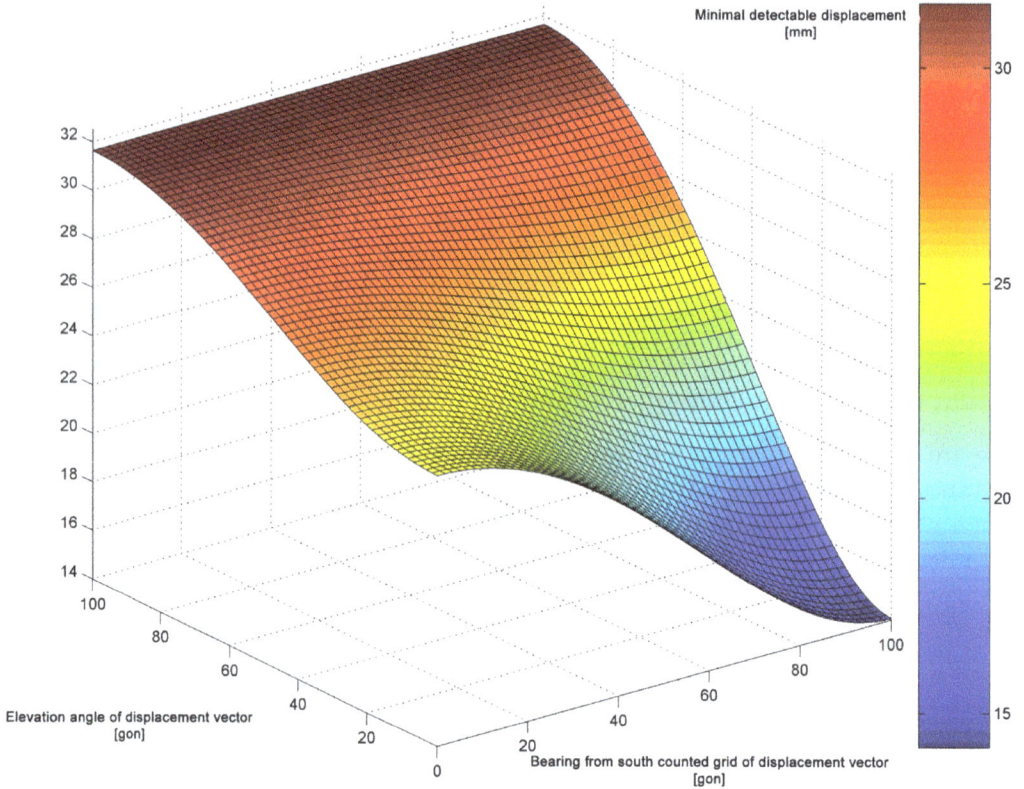

Figure 5: Minimal detectable displacement

Table 2: Unit factors of mean standard deviation in horizontal plane.

Correction Data source	s_{X_0,Y_0}
RTK-FKP	3.18
RTK-PRS	4.63
RTK	3.43

Example (3): Considering values RMS=2.7 mm, HDOP$_1$=HDOP$_2$=1.26, VDOP$_1$=VDOP$_2$ =1.93 and using RTK-FKP service, threshold for reliable displacement detecting would be $\Delta_{\min,H_z} = 15$ mm and $\Delta_{\min,\text{height}} = 38$ mm.

4. Conclusions

It is obvious that a permanent reference network holds great potential for different kinds of geodetic applications, requiring real-time data. The experiment shows some possibilities of land slide monitoring with GPS-RTK technology. Within a short observation time (seconds or a few minutes), it is possible to detect centimetre displacements in real time. This precision is more than sufficient for studying earth surface morphology and its changes [8] as well as for some applications of engineering surveying, such as displacement monitoring. Conclusions of the performed experiment has been used for pilot project of landslide monitoring, in clay quarry "Nepomyšl" [11]. However, as landslide slowed down, the need of higher monitoring precision, unachievable by GNSS/RTK emerged. Based on results of experiments it can be concluded that especially the service RTK-FKP (and its modern version RTK3) is suitable for application of GNSS technologies in the CZEPOS.

Acknowledgements

This contribution was published thanks to the financial support by Ministry of Education, Youth and Sports, the Czech Republic in the Research Program "Udržitelná výstavba" (Sustainable Construction), No. MSM 6840770005 financially supported by Ministry of Education, Youth and Sports, the Czech Republic.

References

[1] Jaroslav Braun and Pavel Hánek. "Geodetic monitoring methods of landslide-prone regions – application to Rabenov". In: *AUC GEOGRAPHICA* 49.1 (Sept. 2014), pp. 5–19. DOI: `10.14712/23361980.2014.2`.

[2] Bernhard Hofmann-Wellenhof, Herbert Lichtenegger, and Elmar Wasle. *GNSS — Global Navigation Satellite Systems: GPS, GLNASS, Galileo and more.* Springer Vienna, 2008, p. 516. DOI: `10.1007/978-3-211-73017-1`.

[3] Jan Kostelecký and Miloš Cimbálník. "Převod souřadnic mezi ETRS-89 a S-JTSK pomocí globálního transformačního klíče (Transformation between ETRS-89 and S-JTSK using Global Transformation Parameters)". In: *Geodetický a kartografický obzor* 53(95).12 (2007).

[4] Jan Kostelecký et al. "Quasigeoids for the Territory of the Czech Republic". In: *Studia Geophysica et Geodaetica* 48.3 (July 2004), pp. 503–518. DOI: `10.1023/b:sgeg.0000037469.70838.39`.

[5] *Land Survey Office. CZEPOS website.* `http://czepos.cuzk.cz/`.

[6] L. Manetti, M. Frapolli, and A. Knecht. "Permanent, autonomus monitoring of landslide movements with GPS". In: *1st Europen conference of landslides. Prague, Czech Republic.* 2002, p. 6.

[7] Leoš Mervat and Miloš Cimbálník. *Vyšší geodézie 2 (Advanced Geodesy 2)*, Czech Technical University in Prague. 178 pp. 1999.

[8] J. Pospíšil and M. Raška. "Geodetic methods in a study of earth surface processes". In: *Geoscape Journal* 1 (2006), pp. 13–20.

[9] M. Raška et al. "Using geodetic techniques for geomorphologic analyses of scree slopes in low-altitude forested regions and its implication for conservation management". In: *Geographia Technica* 2 (2011), pp. 87–100.

[10] G. Retscher. "Accuracy Performance of VRS Networks". In: *Journal of Global Positioning Systems* 1.1 (2002), pp. 40–47.

[11] K. Turčin. "Landslide monitoring in quarry "Nepomyšl"". Mining surveying documentation of Sedlecky kaolin a. s.. (in Czech, company restricted material, unpublished). 2004-2011.

[12]   R. Urban, M. Štroner, and K. Kovařík. "Comparison of GNSS measurement accuracy in reference stations network in territory of Prague". In: *Geodetický a kartografický obzor* 59(101).3 (2013), pp. 45–53.

[13]   J. J. Westerveld. "Mobile Networks, Location Based Services lectures materials". TU Delft, Netherlands.

[14]   G. Wübbena and A. Bagge. *RTCM Message Type 59 – FKP for transmission FKP, version 1.1.* Geo++® GmbH IGS Workshop, Darmstadt, Germany. 8 pp. 2006.

[15]   G. Wübbena, M. Schmitz, and A. Bagge. *Real-Time GNSS Data Transmission Standard RTCM 3.0.* Geo++® GmbH IGS Workshop, Darmstadt, Germany. 26 pp. 2006.

[16]   X. Yang and R. Brock. *RDOP surface for GPS relative positioning.* United States Patent No. 6057800. 2000.

PERMISSIONS

LIST OF CONTRIBUTORS

Eva Novotná
Faculty of Science, Albertov 6, 128 43 Praha 2

Alena Pešková and Jan Holešovský
Department of Geomatics, Faculty of Civil Engineering Czech Technical University in Prague Thákurova 7, 166 29 Prague 6, Czech Republic

Eleni Diamanti and Andreas Georgopoulos
Laboratory of Photogrammetry, National Technical University of Athens

Fotini Vlachaki
Hellenic Institute of Marine Archaeology

Ján Erdélyi, Alojz Kopáčik, Ľubica Ilkovičová, Imrich Lipták and Pavol Kajánek
Slovak University of Technology, Faculty of Civil Engineering, Radlinskeho 11, 813 68 Bratislava, Slovakia

Jiří Šíma
Novorossijská 18, Praha 10, Czech Republic

Michal Med and Petr Souček
The Czech Office for Surveying, Mapping and Cadastre (COSMC) Prague, the Czech Republic

Jan Pacina
J. E. Purkyne University in Ústí nad Labem Czech Republic

Jan Douša, Gabriel Győri and Pavel Václavovic
Research Institute of Geodesy, Topography and Cartography, Geodetic Observatory Pecný Ústecká 98, Zdiby 250 66

Adéla Volfová and Martin Šmejkal
Faculty of Civil Engineering Czech Technical University in Prague

Raffaella Brumana, Daniela Oreni, Mario Alba, Luigi Barazzetti, Branka Cuca and Marco Scaioni
Politecnico di Milano, piazza Leonardo Da Vinci 32, Milan, Italy

Lukáš Brůha
Department of Applied Geoinformatics and Cartography, Charles University in Prague

Peter Löwe
Helmholtz Centre Potsdam GFZ German Research Centre for Geosciences

Tomáš Mildorf and Václav Čada
Department of Mathematics - Section of Geomatics Faculty of Applied Sciences, University of West Bohemia in Pilsen Univerzitní 22, 306 14 Pilsen, Czech Republic

Ikechukwu Maduako
Institute of Geoinformatics, University of Münster, Germany
Director of Studies, Center for Advanced Spatial Technologies & Mapping (CAST-MP) Abuja, Nigeria

Anna Kratochvílová and Václav Petráš
Students of Geoinformatics Programme Faculty of Civil Engineering, Czech Technical University in Prague

Martin Raška
Katastrální úrad pro Karlovarský kraj, Katastrální pracovište Sokolov
Boženy Nemcové 1932, 356 01 Sokolov, Czech Republic

Jiří Pospíšil
Faculty of Civil Engineering, Department of Special Geodesy Thákurova 7, Praha 6 16629, Czech Republic

Index

A

Administrative Units, 43-45, 47, 49

Aerial Laser Scanning, 32, 39, 41

Automation, 122, 129-131, 134, 136-137, 142-144

C

Cadastral Map, 43, 160, 189

Cadastral Parcels, 43-45, 49, 161, 166, 186

Cartographic Documents, 1-2, 4-5, 130, 133

Cartographic Heritage, 5, 130-131, 142

Catalogue System, 134, 136, 144

Clips, 146, 148-156

Cokriging, 81, 88-89, 96, 101-105

Coordinate System, 29, 50, 52-53, 56-57, 59, 135, 195

Cultural Heritage, 24-25, 65, 119-121, 129, 157

Cumulative Solution, 106, 109-110, 112-116

Cz Orthophoto, 32-36

D

Data Integration, 157-158, 162, 165, 169, 172, 184

Database Management System, 135, 172, 179

Datum Transformation, 195-196

Deformation Monitoring, 26-27, 31

Digital Aerial Survey, 32-34, 38, 42

Digital Elevation Model, 32, 37, 39, 89, 165

Digital Surface Model, 20-21, 32, 39

Dtm, 22, 59, 63-65, 119-120, 123, 125

E

Environmental Impact Assessment, 119, 121

Environmental Monitoring, 168-169

Epn-repro1 Project, 106-109, 115

Euref, 67, 69-70, 75-77, 80, 106-107, 112-113, 115, 117-118

G

Geo-web Services, 168, 183

Geobibline, 1-6

Geodetic Control Network, 52-54, 56

Geodetic Monitoring, 26-27, 201

Georeference, 81

Geospatial Data, 32, 81, 135, 172

Geostatistics, 31, 81, 83, 85, 87, 97, 104-105

Gml Files, 43, 45-48

Gnss, 50, 54, 65, 67-70, 76-77, 79, 107-108, 113, 115, 117, 122, 193-194, 201-202

Gnss Correction Signal, 50, 54, 65

Grass, 146-147, 149-156, 185, 187

Gravity Measurements, 7-9

H

Helmert Transformation, 111-113, 142

Heterogeneous Sensor Database, 168-170, 173-176, 178-181, 183

I

In-situ Sensor Observations, 168, 170-172, 178-179

Inspire Directive, 44-47, 121, 133, 157-161

Interpolated Gravity, 7, 9-10

K

Kite Aerial Photography, 50, 55, 65

Knowledge Engineering, 146, 148, 150-152

Kriging, 8, 68, 81, 84-86, 88-89, 96, 99, 103-105

L

Landscape Convention, 119-121, 129

Landslide Monitoring, 193, 200-201

M

Metadata, 1-2, 4-5, 43-46, 48-49, 53, 130-145, 162, 167, 175, 182

Meteorological Data, 67-70, 149

Module G.infer, 146-147, 149-150, 154

Monographs, 1-4

Morphology, 121-123, 196, 200

N

Navigation, 51-54, 79, 106, 117, 123, 189, 201

O

Orthoimagery, 32, 162

Orthophoto Imagery, 32-34

Orthophotomosaic, 20-21
Orthophotomosaics, 15-16, 24

P
Panoramic Views, 119-121, 123-124, 127
Photogrammetry, 15-16, 19, 24-25, 35, 121, 128
Postgis 2.0, 144, 168, 170, 172, 178-179, 183

R
Radiosonde, 67
Raster Data, 81, 130-131, 134-140, 142-143, 147, 168, 172, 175, 184
Reference Frames, 106-107, 116, 118
Reprocessing, 67, 106-108, 115, 117
Rúian, 43

S
Sdi Technologies, 130-131, 142
Sensor Observation Service, 168, 177, 183
Singular Value Decomposition, 26, 30-31
Spatial Data Analysis, 81, 104
Spatial Prediction, 81, 84

T
Technical Metadata, 4, 132-134
Terrestrial Laser Scanning, 26-27, 29-31
Terrestrial Reference Frame, 106-107
Trilateration, 15-18, 20-21
Troposphere, 67-68, 72, 79-80

U
Uav, 65, 119-123, 125-126, 128-129, 168
Underwater Camera Calibration, 15, 19

V
Variogram, 83-88, 94-95, 99-103
Visualization, 15, 71, 84, 103, 114, 125, 134, 136, 138, 141-142

W
Wfs, 43, 46-48, 122, 175-177
Wms, 8, 43, 46-48, 122, 136, 175-177, 189

Z
Zenith Total Delays, 67-68, 79
Zenith Wet Delay, 68, 70